DESIGN AND ANALYSIS OF ALGORITHMS

SECOND EDITION

I. Chandra Mohan

Formerly Professor

Sri Venkateswara University

Tirupati

PHI Learning Private Limited

Delhi-110092

2015

₹ 195.00

DESIGN AND ANALYSIS OF ALGORITHMS, Second Edition
I. Chandra Mohan

ISBN-978-81-203-4575-1

The export rights of this book are vested solely with the publisher.

Third Printing (Second Edition) ... ... February, 2015

Published by Asoke K. Ghosh, PHI Learning Private Limited, Rimjhim House, 111, Patparganj Industrial Estate, Delhi-110092 and Printed by Perfect Impression, New Delhi-110016.

In loving memory of
my parents-in-law
Sri M.N. Rao & Smt Leelavati Rao

Contents

Preface

Based on the feedback received from the readers and students, this thoroughly updated book provides a detailed discussion on the time complexity of algorithms. A few more algorithms are introduced in this editions. Some of the notable features in the form of new insertions are:

Chapter-1: Time complexity of the algorithms for the problem of finding the GCD of two positive integers. Time complexity of "matrix addition" algorithm.

Chapter-2: Time complexity of "matrix multiplication" algorithm. Algorithm and the analysis of time complexity of the "closest-pair elements". Algorithm and computational time bounds for the problem of finding the "Median" of a given sequence of numbers.

Chapter-3: Detailed analysis of the Knapsack problem.

Chapter-7: Analysis of general "Combinatorial search and optimisation" problems.

Chapter-8: Illustration of the "Branch-and-Bound" method with reference to the Knapsack problem.

Chapter-10: The theory of NP-Completeness.

It is hoped with these updates that the second edition will fulfil the aspirations of a large number of readers. The author welcomes any constructive suggestions towards further improvement of the book.

I. Chandra Mohan

Preface to the First Edition

Niklaus Wirth, a famous computer scientist and author of a number of books in Computer Science, entitled one of his books as *Algorithms + Data Structures = Programs*. His intention in coining this title is very clear as it emphasizes the importance of elegant and efficient design of algorithms.

Designing algorithms and analyzing them is an important and primary step in program writing, which is often neglected by program developers. Computer professionals are deeply concerned with devising new and better algorithms for developing efficient program designs. For example, though presently we have a number of sort algorithms, computer scientists are continuously making attempts to design new sort algorithms or modify the existing ones. Is one sort algorithm not enough? Among all the possible algorithms for solving a problem, we look for an "efficient algorithm". The factors that make one algorithm more efficient than other algorithms are discussed in this book. Thus, designing an algorithm and analyzing it are equally important issues.

This book is suitable for a course on "Design and Analysis of Algorithms", which is offered as a core paper to B.Tech students (CSE, ECE, EIE), undergraduate and postgraduate students of computer science and computer applications (BCA/MCA, BSc(CS)/MSc(CS)), and other computer-related courses in Information Technology. It is a student-friendly book whose contents are designed to be covered in a one-semester period.

The author welcomes comments and suggestions for improving the quality of the contents of this book.

I. Chandra Mohan

Acknowledgements

Among the many available books on the subject, I have a special appreciation for the book *Fundamentals of Computer Algorithms* and a lot of praise for its authors—Ellis Horowitz, Sartaj Sahni and Sanguthevar Rajasekaran. Having followed this book in teaching for a number of years, it is natural that it has had its own influence at least on the structure and sequence of topics in presenting the subject material in my book. I profoundly thank the three authors for enlightened learning of the subject by me, from their book.

I am thankful to my wife Pramila for the constant support and encouragement I received from her and also for making necessary corrections in the manuscript. I am grateful to my daughter Padma Suhasini and my son-in-law Vijaya Kumar for their moral support in writing this book.

Finally, I am grateful to many people at PHI Learning who worked on this project. Their efforts are much appreciated.

I. Chandra Mohan

CHAPTER 1

Fundamentals

1.1 INTRODUCTION

An algorithm is a tool for solving a given problem. An algorithm is a finite sequence of instructions such that each instruction:

1. has a well defined and clear meaning;
2. can be performed in a finite amount of time; and
3. can be performed easily.

Before writing a code (program) for solving the given problem, a good programmer first designs and writes the concerned algorithm, analyses it and refines it as many times as required and arrives at the final 'efficient' form that works well for all valid input data and solves the problem in the shortest possible time, utilising minimum memory space.

Definition [Algorithm]: An algorithm is a logically-arranged sequence of a finite number of instructions which when implemented solves the given problem.

It is commonly agreed upon by the computer scientists that all algorithms have the following characteristics: (1) Input, (2) Output, (3) Definiteness, (4) Finiteness, and (5) Effectiveness.

There can be no input or any number of inputs but at least one output must be generated. There should not be any confusion in any instruction and each instruction must be specific and unambiguous. We expect that each instruction and thereby the entire algorithm terminates after a finite amount of time, for all possible valid input data. Each instruction must be effective. It must be possible for any person to check the effectiveness of the algorithm with some sample data. The operations involved must not only be definite but also feasible.

1.2 ALGORITHM ANALYSIS

Following are some of the basic factors in designing and analysing algorithms:

(a) **Designing algorithms:** This part consists of designing and writing algorithms for solving problems. Designing of algorithms is done by using different types of methods. The commonly used methods are: divide-and-conquer, dynamic programming, branch-and-bound, backtracking, and greedy methods.

(b) **Validation of algorithms:** First, we design the required algorithm and check whether it computes the correct answer for all possible valid inputs. This part of checking is called **algorithm validation**. It has to be proved that the program is correct. This proof involves establishing that each statement in itself is well defined as well as the correctness of all basic operations.

(c) **Analysing an algorithm:** The analysis of algorithm is concerned with the estimation of the resources it requires. But of all these estimates, the total time that the algorithm requires to complete implementation of all of its instructions is taken as a measure of performance of the algorithm. For such estimates, we use a generic-processor model of computations, called the **random access machine (RAM)**. If we write two algorithms A and B for solving the same problem, each having time complexities $T(A)$ and $T(B)$ respectively, and if $T(A) < T(B)$, then we prefer algorithm A to that of B.

(d) **Testing a program:** Write a program and execute it with some sample data and check whether you get an output. Otherwise, you make necessary corrections in the program and execute it again. This process continues till you get output. This part is called "**debugging**". However, it does not mean that the program works well for all other legally valid input data sets. A better way is to allow more programmers to write their own program independently for the same problem and then execute each program for the same input data. If the outputs are same, then we can say with confidence that they are correct. By executing a correct program on a number of legally valid data sets one can measure the time and space complexities for its computation.

1.3 SOME SAMPLE ALGORITHMS

In this section we present two sample algorithms.

ALGORITHM: BUBBLE SORT

An array "a" of size n of unordered elements is given. "Bubble Sort" algorithm sorts the elements of the array "a" into either ascending or descending order. Note that this algorithm requires at most $(n - 1)$ passes. In the following algorithm, the variables "count" and "loc" denote the counter and location of the recent unsorted element respectively. The variable "num" gives the number of exchanges made in any pass. Refer to algorithm 1.1.

Algorithm 1.1 Algorithm for the Bubble Sort problem

```
Algorithm BubbleSort (A, n);
// This algorithm sorts the array A[1:n].
{
     loc := n;
   for i:=1 to n-1 do
     {
          num := 0;
   // Compare pairwise unsorted elements
          for i := 1 to loc - 1 do
            if a[i] > a[i+1] then // interchange a[i] and a[i+1]
              {
                temp := a[i];
                a[i] := a[i+1];
                a[i+1] := temp;
              }
          num := num + 1;
            if num = 0 then // Check whether any exchanges made
              return (1);
            else
              loc := loc - 1;
     }
          return (num);
 }
```

ALGORITHM: TOWERS OF HANOI

This is the best example of recursion. Three vertical towers (rods) labeled A, B and C are fixed erect on a platform. On tower A, there are n disks. The disks were of decreasing size (diameter). The largest disk at the bottom, over it the next largest and finally with the smallest-sized disk at the top. The problem is to move all the n disks on tower A to tower B maintaining the same order of size (bottom to top). Tower C can be used for intermediate movement of disks. Suppose $n = 3$ and let the largest disk be D_1 and the next largest disk be D_2 and let the smallest disk be D_3. The following sequence of steps solves this problem.

1. Move D_3 to tower B.
2. Move D_2 to tower C.
3. Move D_3 which is now on tower B to tower C.
4. Move D_1 to tower B.
5. Move D_3 which is now on tower C to tower A.
6. Move D_2 which is on tower C to tower B.
7. Finally, move D_3 which is now on tower A to tower B.

Observe that the number of movements when $n = 3$ is $(7 = 2^3 - 1)$. In general, this number is $(2^n - 1)$. When $n = 64$, think of the size of this number.

Suppose the top $(n - 1)$ disks are moved to tower C maintaining the order of size (bottom to top). Then move the single left out bottommost (largest) disk on tower A to tower B. Move the $(n - 1)$ disks from tower C to tower B. This is recursion. The n disk problem is divided into two $(n - 1)$ disk problems.

The concerned algorithm is given below.

Algorithm 1.2 Algorithm for the Towers of Hanoi problem

```
Algorithm HanoiTowers (n, A, B, C)
// This algorithm moves the top n disks from tower A to tower B.
{ if ( n ≥ 1) then
        {
        HanoiTowers (n-1, A, C, B);
  write ("Move disk on the top of tower", A, "to top of tower", B);
   HanoiTowers (n-1, C, B, A); } }
```

1.4 SPACE COMPLEXITY

The space complexity of an algorithm P is the amount of memory it requires for running the algorithm. It has two parts, viz. a fixed part and a variable part. Thus

$$S(\text{P}) = c + S_{\text{P}} \text{ (instance characteristics)}$$

where "c" is a constant.

The fixed part component c, is independent of the characteristics of inputs and outputs and depends on the space for the code and space for certain type of variables. The variable part component S_{P} depends upon the space needed by the component variables, referenced variables and the recursion stack space. This recursion stack space includes the space for the formal parameters, local variables and the return address.

We estimate S_{P} for a given program P and this method is illustrated by the following two simple algorithms.

Let us consider the following iterative algorithm for finding the sum of elements of an array:

Algorithm 1.3 An iterative algorithm for the sum and product of elements of an array

```
AlgorithmTotProd (a, n)
{
     tot := 0.0 ;
  prod : = 1.0;
   for k :=1 to n do
        {
```

```
            tot := tot + a[k];
            prod := prod * a[k];
        }
    return tot;
    return prod;
}
```

In this algorithm the space needed by n, i and tot and prod is one word for each and for the array a [ ] is at least n words. Thus

$$S_P = S_{\text{ArrayTot}}\ (n) \geq n + 4$$

Consider the following recursive algorithm for the above problem:

Algorithm 1.4 Array elements product — A recursive algorithm

```
Algorithm RProd (a , n)
  {
    if (n = 1) then
    return a[1];
    else
      return RProd (a, n-1) * a[n];
}
```

In this algorithm, the instances are characterized by n. Let us suppose that the return address requires one word of memory. At least three words (space for n, the return address and a pointer to a[]) are required for each call to RProd. Thus, the recursion stack space is greater than or equal to $3(n+1)$, as depth of recursion is $(n + 1)$.

1.5 TIME COMPLEXITY

With the execution of every program the following two types of times are associated: (1) the compile time which does not depend upon the instance characteristics and (2) the run time of the program. The time T(P) taken by a program P is the sum of the compile time and the run time. However, we give importance to the run time part only for the simple reason in addition to its aforesaid characteristic, that a program that is already compiled can be executed any number of times without being compiled again and again. We denote the run time of a program P by rt_P (instance characteristics). rt_P depends upon a number of factors which vary from one program to another. So, we estimate rt_p. If n number of characteristics are involved in P and if T_i is the time taken for the ith characteristic CH_i, then $rt_p(n)$ is the sum of products of T_i and $CH_i(n)$. Some frequently occurring characteristics are additions, multiplications, subtractions, comparisons.

"A **program step** is defined as syntactically and semantically valid statement of a program whose run time that is independent of instance characteristics". The step sizes in a program are computed as follows:

1. Step size of a comment is zero.
2. The step size of an assignment statement which does not call any other algorithms is taken as one.
3. In iterative statements like **for, while, repeat, until,** the step count for each execution of the control part is incremented by one.

One way of measuring the final (total) step size of a program is to introduce a new variable, say "count" and increment it by an appropriate quantity whenever a statement is executed.

The following algorithm is a variation of algorithm 1.3 in which statements that count the number of steps in the algorithm 1.3 alone are shown. For this purpose a variable "num" is used.

Algorithm 1.5 Algorithm 1.3 with count statement

```
1.  Algorithm ATotProd (a, n)
2.  //The variable num is the count which is global and is initialized
    as zero.
3.  {
4.      tot := 0.0;
5.      prod :=1.0;
6.      num :=num+2;
7.      for k :=1 to n do
8.      {
9.          num := num+1;
10.         tot := tot + a[k];
11.         Prod :=Prod * a[k];
12.         num:=num+1
13.     }
14.     num := num+1;
15.     return tot;
16.     return prod;
17.     num := num+2;
18. }
```

The variable num is first initialized by zero. In statement 6, we increment by one in view of the assignment statement 4 and 5. Similarly, in statement 9, it is for the "for loop", in statement 12 it is for the two assignment statements 10 and 11, in statement 12, in statement 14 the increment is for the it is for the last execution of the "for loop" and finally in statement 17 it is for the return statements 15 and 16

Algorithm 1.6 Algorithm 1.4 with count statement

```
1.  Algorithm RArrayProd (a, n)
2.  {
```

```
    num := num+1;
    if (n = 0) then
     {
       num := num+1; return a [1];
     }
    else
     {
      num := num+1; return RArrayProd (a, n-1)*a[n];
  } }
```

Now we discuss the run time $rt_{\text{RArrayProd}}(n)$, for the recursive algorithm 1.6. Clearly, $rt_{\text{RArrayProd}}(0) = 2$. When $n > 0$, count is increased by 2 and also by $rt_{\text{RArrayProd}}(n-1)$ in view of calling RArrayProd from the **else** part.

Thus

$$rt_{\text{RArrayProd}}(n) = \begin{cases} 2 & \text{if } n = 0 \\ 2 + t_{\text{RArrayProd}(n-1)} & \text{if } n > 0 \end{cases}$$

We can solve the above recurrence relation as follows, by repeated application of it:

$$\begin{aligned} rt_{\text{RArrayProd}}(n) &= 2 + rt_{\text{RArrayProd}}(n-1) \\ &= 2 + 2 + rt_{\text{RArrayProd}}(n-2) \\ &= 2(2) + rt_{\text{RArrayProd}}(n-2) \\ &= \cdots\cdots\cdots \\ &= 2n + rt_{\text{RArrayProd}}(0) = 2n + 2. \end{aligned}$$

1.6 INPUT SIZE

Input size is an instance characteristic. It shows the number of words required by that instant characteristic.

Now we illustrate the process of counting the number of steps with the following two algorithms for the addition of two $m \times n$ matrices X, Y to give the matrix Z:

Algorithm 1.7 Algorithm for matrix addition with counting statement

```
Algorithm MatSum(X, Y, Z, m, n)
  {
     for j := 1 to m do
        {
          num := num + 1;
           for k := 1 to n do
```

```
            {
                num := num + 1 ;
                Z[j, k] := X[j, k] + Y[j, k];
                num := num + 1;
                }
                num := num+1;
        }
        num := num+1;
    }
```

Algorithm 1.8 A simplified algorithm of algorithm 1.7 with count statement only

```
Algorithm Add (X, Y, Z, m, n)
{
    for j := 1 to m do
    {
        num := num+2;
        for k := 1 to n do
        num := num+2; }
    num := num+1; }
```

We see that the step count in algorithm 1.8 is $2(mn + m) + 1$ and the input size is $2mn + 2$.

Another way to determine the step count of any algorithm is to build a table that shows the number of steps per execution (s/e) of a statement and the total step size. Then the step count for the algorithm is obtained.

The s/e of a statement is the amount by which the count changes as a result of the execution of that statement.

The table for algorithm 1.3 is given below:

Table 1.1 Step table for summation of elements of an array

Statement #	*s/e*	*Frequency*	*Total steps*
1	0	—	0
2	—	—	—
3	1	1	1
4	1	1	1
5	1	$n + 1$	$n + 1$
6	—	—	—
7	1	n	n
8	1	n	n
9	—	—	—
10	1	1	1
11	1	1	1
12	—	—	—
Total			$3n + 5$

Following is an algorithm for multiplying three $n \times n$ matrices.

Algorithm 1.9 Algorithm for matrix multiplication

```
Algorithm MatMul (a, b, c,d)
//This algorithm multiplies the two n×n matrices and stores
   the resultant
//matrix in c.
   {
      for i := 1 to n do
         for j := 1 to n do
            {
              c[i, j] :=0.0;
              for k :=1 to n do
                c[i, j] : = c[i, j] + a[i,k]*b[k, j];
         }
   }
   return c;
}
```

Following is the step table for the above recursive algorithm 1.4. (Assuming that $n > 0$).

Table 1.2 Step table for the recursive algorithm 1.4

Statement #	*s/e*	*Frequency*	*Total steps*
1	0	—	0
2	—	—	—
3	1	1	1
4	1	0	0
5	—	—	—
6	$1 + rt_{\text{RArrayTot}}(n-1)$	1	$1 + rt_{\text{RArrayTot}}(n-1)$
7	—	—	—
Total			$2 + rt_{\text{RArrayTot}}(n-1)$

EXERCISES

1.1 Write a simple algorithm for matrix addition without any count statements and construct the step table for that algorithm.

1.2 Write an algorithm for generating n (>1) Fibonacci numbers and construct the step table for it.

We now discuss the **problem of finding the GCD** of two positive integers m, n and elaborate the basic two algorithms in this context and carry out the analysis of their time complexities.

We suppose that $m < n$. Otherwise, we first interchange them. The greatest common divisor of m, n is denoted by GCD(m, n). The first method is very simple and straightforward and this method is called "factorization method." Factorise m and n as the product of primes. Let

$$m = m_1 * m_2 * \cdots * m_k$$

and

$$n = n_1 * n_2 \cdots * n_l$$

where each of m_i and n_j is prime.

The product of common factors in m and n gives GCD(m, n).

Suppose $m = 72$ and $n = 96$.

Then it is clear that

$$m = 2 * 2 * 2 * 3 * 3$$

and

$$n = 2 * 2 * 2 * 2 * 2 * 3$$

The product of common factors in both m and n is: $2 * 2 * 2 * 3 = 24$.

Thus the GCD(72, 96) = 24. Note that in this problem, we used 11 divisions.

The other one is to use the well-established **Euclid's algorithm**. The crucial steps in Euclid's algorithm are as follows.

```
Euclid
{
    while m does not divide n
            r = n mod m ;
            n = m ;
            m = r ;
    end     /* end of while loop */
}
```

We illustrate the working of the Euclid's algorithm for the same aforesaid numbers: $m = 72$ and $n = 96$.

```
Iteration-1:        Does m divide n? NO      /* Enter the while loop */
                    r = 96 mod 72 = 24
                    n = 72
                    m = 24
Iteration-2:        Does m divide n ? YES
```

The while loop terminates and the value 24 is returned as GCD of 72 and 96. Note that only 2 divisions made this algorithm.

This algorithm seems to be good. This fact is to be established, which we have shown below.

Note that if m divides n, then GCD(m, n) = m, otherwise

$$\text{GCD}(m, n) = \text{GCD}((n \bmod m), m).$$

Suppose that GCD(m, n) = k. Then we can write m and n as

$$m = p * k \text{ and } n = q * k$$

where p and q are relatively prime to each other.

Notice that (n mod m) $< m$. Thus as we do one iteration after another, the values of m, n become smaller than their previous values. But, the question is how smaller the values of m and n can be. The two numbers are always positive integers. Either of them cannot be zero. Thus the loop terminates after some finite number of iterations. Also, notice that at every iteration we compute the GCD of new numbers but the GCD always remains the same as the GCD(m, n). In this connection, we have the following result:

Result: At every iteration of Euclid's algorithm for finding GCD of two integers, p and q, their values get reduced to smaller values, say m and n respectively. It can be shown that the sum of the values m and n, at each iteration decreases by 3/2 of their previous iterated values. Values p and q are called instances of the values m and n, respectively.

At the first iteration

$$m = p$$

and

$$n = q$$

Let at the second iteration

$$m = p_1$$

and

$$n = q_1$$

Consider the ratio $(p + q)/(p_1 + q_1)$.

We observe that the new value of n is the old value of m and the new value of m is n mod m. So

$$q_1 = p$$

and

$$p_1 = q \bmod p$$

We see that

$$p_1 < p$$

and

$$\begin{aligned} p_1 + q_1 &= \text{Remainder} + \text{divisor} \\ &\leq \text{dividend} \end{aligned}$$

Here the dividend is q. So

$$p_1 + q_1 \leq q$$

Now consider the expression: $p_1 + q_1 + 2\,(p_1 + q_1)$

In view of the above argument,

$$p_1 + q_1 + 2\,(p_1 + q_1) < p + p + 2q$$

That is

$$3\,(p_1 + q_1) < 2(p + q)$$

Thus $$(p_1 + q_1) < (p + q)/(3/2)$$

Thus, the sum of values of the variables m and n decreases by a factor 3/2.

This implies

Number of iterations $\leq \log_{3/2}(p + q)$

Now consider the **matrix addition problem**. Without bothering about the minor details such as input/output statements, dimension declaration and initialization statements, we consider the following crucial step in the matrix addition problem.

```
for i = 1 to n
    C[i, j] := A[i, j] + B[i, j] ;
```

The actual and detailed implementation of the above loop is:

```
Step number        Statement
    1.          i = 1
    2.                    /* check */
                if i > n then go to the end of the loop
                else
                          /* body of the loop begins */
    3.      x = A[i]
    4.      y = B[i]
    5.      sum = x + y
    6.      C[i] = sum
            /* body of the loop ends */
    7.      i = i + 1
    8.      go to step 2
    9.      End of the for loop
```

Step 1 is executed one time. Step 2 is executed $(n + 1)$ times. Suppose, in general, the body of the *for loop* contains b number of instructions. Then the number of times all these instructions get executed is bn. Steps 7 and 8 are executed n times each. The number of times all steps are executed by the end of the for loop is

$$1 + (n + 1) + bn + n + n$$

$$= 2 + n(b + 3)$$

The time complexity of this algorithm depends upon $n(b + 3) + 2$

1.7 ASYMPTOTIC NOTATION

We use the following notation, called the asymptotic notation used in analysing algorithms.

Definition [Big "oh"]: Let $f(n)$ and $g(n)$ be two functions. Then we say $f(n)$ is big oh of $g(n)$ and write $f(n) = O(g(n))$ if and only if there exists constants c and n_0, such that

$$f(n) \leq c * g(n) \qquad \text{for all } n \geq n_0$$

Examples

1. $3n + 1 = O(n)$, as $3n + 1 \leq 4n$ for all $n \geq 1$.
2. $3n + 4 = O(n)$ as $3n + 4 \leq 5n$ for all $n \geq 2$.

3. $5*2^n + n^2 = O(2^n)$ as $5*2^n + n^2 \le 7*2^n$, for all $n \ge 1$.
4. $10 * n^2 + 4*n + 2 = O(10 * n^4)$ as $10 * n^2 + 4*n + 2 \ge 10 * n^4$ for all $n \ge 2$.
5. $3n * n^4 + 2 \ne O(1)$ as $3n + 2$ is not $\le c$ for any constant.

The computing times having the following orders are referred frequently: $O(1)$ (= constant), $O(\log n)$, $O(n)$, $O(n \log n)$, $O(n^2)$, $O(n^3)$ and $O(2^n)$.

$f(n) = O(g(n))$ means that $g(n)$ is an upper bound on the value of $f(n)$ for all $n \ge n_0$, for some positive integer n_0.

$g(n)$ that is selected may not be a good upper bound for $f(n)$ for all $n \ge n_0$ but it is an upper bound. It is suggested to take $g(n)$ as small as possible to have a better upper bound for $f(n)$.

Definition [Little "oh"]: We say $f(n)$ is little oh of $g(n)$ and write $f(n) = o(g(n))$ if and only if limit of $(f(n)/g(n))$ as $n \to \infty$ is zero.

Examples

6. $3n + 2 = o(n^2)$
7. $3n + 2 = o(n \log \log n)$
8. $6 * 2^n + n^2 = o(3^n)$
9. $6 * 2^n + n^2 = o(2^n \log n)$

Definition [Omega]: We say $f(n)$ is omega of $g(n)$ and write $f(n) = \Omega(g(n))$ if and only if there exists positive constants c and n_0, such that

$$f(n) \ge c * g(n) \qquad \text{for all } n \ge n_0$$

Examples

10. $3n + 2 = \Omega(n)$ as $3n + 2 \ge 3n$ for all $n \ge 1$.
11. $10n^2 + 3n + 1 = \Omega(n^2)$ as $10n^2 + 3n + 1 \ge 10n^4$ for all $n \ge 1$.
12. $5*2^n + n^2 = \Omega(2^n)$ as $5*2^n + n^2 \ge 2^n$ for all $n \ge 1$.

Consider the Fibonacci series defined by the recurrence relation

$$F(n) = F(n-1) + F(n-2), \text{ with } F(1) = F(0) = 0$$

It is easy to establish that $F(n) \ge 2^{n/2}$ and hence

$$F(n) = \Omega\,(2^{n/2})$$

When $f(n) = \Omega(g(n))$, then $g(n)$ is a lower bound on $f(n)$. We select $g(n)$ as large as possible to get a better upper bound for $f(n)$.

Definition [Theta]: We say $f(n)$ is theta of $g(n)$ and write $f(n) = \Theta(g(n))$ if and only if there exists positive constants c_1 and c_2, and n_0 such that

$$c_1 g(n) \le f(n) \le c_2 g(n) \qquad \text{for all } n \ge n_0.$$

Examples

13. $3n + 2 = \Theta(n)$ as $3n \le 3n + 2 \le 4n$, for all $n \ge 2$ and for $c_1 = 3$ and $c_2 = 4$
14. $3n + 3 = \Theta(n)$
15. $10n^2 + 4n + 2 = \Theta(n^2)$

Consider the functions

$$f(n) = 2 + \frac{1}{n}$$

and

$$g(n) = 1$$

We observe that

$$2 \le 2 + \frac{1}{n} \le 3$$

Thus

$$c_1\, g(n) \le f(n) \le c_2\, g(n) \text{ with } c_1 = 2,\ c_2 = 3 \text{ and } n_0 = 1$$

Hence, in this example

$$f(n) = \Theta(g(n))$$

Note: As defined earlier, $f(n) = \Theta\,(g(n))$ if there exists constants c_1 and c_2 such that

$$c_1\, g(n) \le f(n) \le c_2\, g(n)$$

for all $n \ge n_0$.

This implies that the function will always lie between or equal to $c_1 g(n)$ and $c_2 g(n)$ for all $n \ge n_0$. The shape of the function is immaterial for other values of n. We also note the following interesting features of Θ.

If $f(n) \in \Theta\,(g_1(n))$ and $h(n) \in \Theta\,(g_2(n))$ then

$$f(n) + h(n) \in \Theta\,(g_1(n) + g_2(n))$$

and if $g_1(n) = g_2(n)$, then $f(n) + h(n) \in \Theta\,(g(n))$.

1.8 PERFORMANCE ANALYSIS OF ALGORITHMS

Performance analysis of an algorithm means the investigations of the behaviour of the algorithm for different inputs and for different sizes of inputs and also its asymptotic behaviour.

QUESTIONS

1. Define an algorithm and discuss the issues pertaining to algorithms. Write the algorithm for the "Tower of Hanoi" problem.
2. Define space and time complexities of an algorithm. Write the selection sort algorithm and discuss its space and time complexities.
3. What is a step table? Write an algorithm for matrix addition problem and give its step table.

4. Write a non-recursive algorithm for generating "Fibonacci Numbers" and discuss its time complexity.
5. What is a recurrence relation? Solve the following recurrence relations.

(a) $T(n) = \begin{cases} a & \text{if } n = 1 \\ 2T(n/2) + cn & \text{if } n = 2^k \text{ for some positive integer } k. \end{cases}$

(b) $T(n) = \begin{cases} 2T\ n/2 + 2 & \text{if } n > 2 \\ 1 & \text{if } n = 2 \\ 0 & \text{if } n = 0. \end{cases}$

6. Define input size. Write an algorithm for matrix addition and construct step table for it.
7. Explain with examples Big oh, Omega, Theta, and Little oh asymptotic notations.

CHAPTER 2

Divide-and-Conquer Methods

2.1 INTRODUCTION

The divide-and-conquer methods are used in designing algorithms when the function to be computed has n inputs. As the title of this method indicates, the original problem of n inputs is divided, recursively by reapplication of the divide-and-conquer principle, into m sub-problems for $1 \leq m \leq n$ till it becomes easy to solve each of these m sub-problems individually. We then obtain m sub-solutions. Finally, a suitable method is devised to unite all these m sub-solutions to obtain a solution of the original problem. We insist that all the sub-problems be of the same type as the original problem.

The divide-and-conquer paradigm consists of the following three steps at each level of recursion:

1. Divide the problem into a number of sub-problems,
2. Conquer the sub-problems by solving them and
3. Combine the solutions of all the sub-problems to obtain a solution for the original problem.

Let P be the original problem. We consider a boolean function Small(P) that output will be "true" when the input size is small. Then the solution of the problem can be obtained directly without further splitting. If Small(P) returns a "false", then the problem P is divided into smaller sub-problems.

In general, the mathematical model, of the solution to a problem when divide-and-conquer technique is applied, turns out to be a recurrence relation. By solving this recurrence relation, we obtain the solution to the original problem. Some methods of solving recurrence relations are discussed later in this chapter.

2.2 SEARCH AND SORT ALGORITHMS

In this section, we apply the divide-and-conquer technique in solving a "search" problem and some "sort problems".

2.2.1 Binary Search

We are given a sorted array $a[1 : n]$ of size n and a search element x. The problem is to determine whether x is present in the list $a[\,]$ or not. If it is present, we determine a value j such that $a[j] = x$, $1 \le j \le n$. If x is not present in the list, then a proper action will be taken, may be one can set conveniently, $j = 0$.

An instance of the problem depends upon n, some contiguous elements in the array $a[\,]$ and search element x. We take an arbitrary instance P = $(n, a_i, \ldots, a_k, x)$ for some $k(\le n)$ for this problem. S(P) will receive the value i if $x = a[i]$, otherwise zero. Let $g(n)$ be the time to solve the problem for small inputs. Suppose we assign "true" to Small (P) when $n = 1$. Then S(P) takes the value j such that $a[j] = x$. Otherwise, S(P) takes the value 0. Then $g(1) = \Theta(1)$. P split into smaller problems if P has more than one input. We compare x with $a[m]$, for some m such that $i \le m \le k$. If $x = a[m]$, then the search is successful and the problem is solved and $j = m$. If $x < a[m]$, then $a[i] \le x \le a[m-1]$. So x has to be searched only in the sub-list $a[i]$ $a[i + 1], \ldots, a[m-1]$ and P reduces to $(m - i, a[i], \ldots, a[m-1], x)$. The only alternative left out is that $x > a[m]$. In this case, $a[m + 1] \le x \le a[k]$. x has to be searched in the sub-list $a[m + 1], \ldots, a[k]$. P now reduces to $(k - m, a[m + 1], \ldots, a[k], x)$.

Division of P into smaller subproblems take only $\Theta(1)$ time. We apply the divide-and-conquer method again and again to solve the instance, if necessary.

This divide-and-conquer method is called a binary search method if m is chosen every time such that a_m is the middle element of the list, i.e., $m = \lfloor (n + 1)/2 \rfloor$.

Now we represent recursive and iterative algorithms 2.1 and 2.2 for the binary search.

Algorithm 2.1 Recursive binary search

```
Algorithm BinRSearch(a, j, k, x)
// a[j, k] is a given sorted array of non-decreasing elements
// order, 1 ≤ j ≤ k. This algorithm searches the list for the
// presence of x . If the search is successful, i.e. if x is
// present, it returns i such that x = a[i], else returns zero.
{
   if ( k = j ) then // Small(P)
      {
         if ( x = a[j] ) then return j;
           else return 0;
```

```
   }
   else
   {// P is divided into smaller subproblems.
   mid : = ⌊(j + k) /2⌋
      if ( x = a[mid] then return mid;
      else if ( x < a[mid] then
         return BinRSearch(a, j, mid - 1, x);
         else return BinRSearch(a, mid + 1, k, x);
   }
}
```

Algorithm 2.2 Iterative binary search

```
Algorithm ItBinSearch(a,n,x)
// a[1:n] is an array of n elements in non- decreasing
// order. This algorithm determines whether x is present in
// a[ ] and if so returns i such that x = a[ i ], else returns 0
{
   l:=1; h:=n;
   while(l d” hi) do
     {
        mid:= ⌊(l + h)⌋/2;
        if (x < a[mid]) then
        h:=mid-1;
           else if (x > a[mid]) then
              l :=mid+1;
           else return mid ;
   }
   return 0;
}
```

Time complexity of binary search

We observe that for the binary search problem, the time complexity $T(n)$ is defined by the recurrence relation

$$T(1) = 1$$
$$T(n) = T(n/2) + 1, \quad n \geq 1$$

Now we solve this recurrence relation

$$\begin{aligned} T(n) &= (T(n/4) + 1) + 1 = T(n/4) + 2 \\ &= (T(n/8) + 1) + 1 + 1 \\ &= (T(n/8) + 3) \end{aligned}$$

$$= \ldots \ldots \ldots \ldots \ldots$$

$$= T(n/2^k) + k$$

Now let $2^k = n$, i.e., $k = \log_2{}^n$. Then, $T(n) = T(1) + \log_2{}^n = 1 + \log_2{}^n$.

$\therefore$ Time complexity of binary search $= O(\log_2{}^n)$.

	Successful searches	*Unsuccessful searches*
Best:	$\Theta(1)$	$\Theta(\log n)$
Average:	$\Theta(\log(n))$	$\Theta(\log n)$
Worst:	$\Theta(\log n)$	$\Theta(\log n)$

Exercise

2.1 Apply the above two binary search algorithms for the data items: 8, 10, 24, 55, 83, 102, 113, 126, 132, 143 and 152.

2.2.2 Merge Sort

We apply the divide-and-conquer method for solving the "Merge Sort" problem. The welcoming feature of this method is that the worst case time complexity is just $O(n \log_2 n)$ only and there is less movement of data. We split the given unsorted array $a[1:n]$ of elements n, into two sub-arrays, say into $a[1:k]$ and $a[k+1:n]$ (normally, k is taken as the mid value between 1 and n). These sub-arrays are further split recursively into smaller and smaller sub-arrays till each sub-array can be sorted directly. Then we combine (merge) them.

Algorithm 2.3 illustrates the "Merge Sort" technique.

Algorithm 2.3 A Merge Sort algorithm

```
Algorithm MergeSort(i, j)
// a[i, j] is an unsorted array.
// Small(P) is true if P can be solved directly.
{
      if (i < j) then // S(P) is false
      {
         // Divide P into smaller sub-problems
         mid := ⌊(i + j)⌋
            MergeSort(i, mid);
            MergeSort(mid+1, j);
         // Combine the sub-solutions
            Merge(i, j, mid);
} }
```

Algorithm 2.4 presents the merging of the sub solutions.

Algorithm 2.4 Merging two sorted arrays

```
Algorithm Merge(i, j, mid)
// This algorithm combines two sorted arrays
// a[i : mid] and a[mid+1 : j] into a single sorted array a[i : j]
{
    save: = i;
    p: = i;
    q: = mid + 1;
  while (( save d" mid) and (q ≤ j)) do
  {
  if (a[save] ≤ a[q]) then
  {
  b[p]: = a[save];
  save: = save + 1;
  }
  else
  {
    b[p]: = a[q];
    q: = q + 1;
  }
  p: = p + 1;
  }
  if (save > mid) then
    for k: = q to j do
  {
  b[p] := a[k];
  p: = p + 1;
  }
  else
    for k: = save to mid do
  {
    b[p]: = a[k];
    p:= p + 1;
  }
      for k: = i to j do
        a[k]: = b[k];
}
```

Example 2.1

Trace the above algorithm for the sample data: a[1:10] = {315, 290, 184, 657, 356, 428, 866, 259, 455, 525}.

The above algorithm, splits the original array into two sub arrays $a[1:5]$ and $a[6:10]$. The elements in $a[1:5]$ are then split into two sub-arrays $a[1:3]$ and $a[4:5]$. Then $a[1:3]$ is split into the two sub-arrays $a[1:2]$ and $a[3:3]$. The two elements in $a[1:2]$ are split into the one-element sub-arrays $a[1]$ and $a[2]$. We now merge the sub-arrays. Different steps involved in splitting the given array are shown below.

(22 | 20 | 8 | 67, 35 | 44, 75, 15, 53, 59)

Elements $a[1]$ and $a[2]$ are merged giving

(20, 22 | 8 | 67, 35 | 44, 75, 15, 53, 59)

Then $a[3]$ is merged with $a[1:2]$ giving

(8, 20, 22 | 67, 35 | 44, 75, 15, 53, 59)

Next the elements a[4] and a[5] are merged:

(8, 20, 22 | 35, 67 | 44, 75, 15, 53, 59)

and then $a[1:3]$ and $a[4:5]$

(8, 20, 22, 35, 67 | 44, 75, 15, 53, 59)

Now the second recursive call for MergeSort begins. Repeated application of this gives:

(8, 20, 22, 35, 67 | 44 | 75 | 15| 53, 59)

Elements a[6] and a[7] are merged.

Then $a[8]$ is merged with $a[6:7]$ to give

(8, 20, 22, 35, 67 | 15, 44, 75 | 53, 59)

Now a[9] and a[10] and then a[6 : 8] and a[9 : 10] are merged to give:

(8, 20, 22, 35, 67 | 15, 44, 53, 59, 75)

We now have two sorted sub-arrays. The final merge gives the sorted list

(8, 15, 20, 22, 35, 428, 44, 59, 67, 75)

We now compute the time complexity $T(n)$ of this MergeSort algorithm. We suppose that the time taken for merging is proportional to n. Then

$$T(n) = \begin{cases} a & \text{if } n = 1, \text{`}a\text{' being a constant} \\ 2T(n/2) + cn & \text{if } n > 1, \text{`}c\text{' being a constant} \end{cases}$$

When n is a power of 2, $n = 2^k$, we can solve this equation by the method of successive substitutions:

$$\begin{aligned}T(n) &= 2(2T(n/4) + cn) + cn \\ &= 4T(n/4) + 2cn \\ &= 8T(n/8) + 3cn \\ &\vdots \\ &= 2^K T(1) + Kcn \\ &= an + cn \log n\end{aligned}$$

If $2^k < n < 2^{k+1}$, then $T(n) \le T(2^{k+1})$.

Therefore, $T(n) = O(n \log n)$.

2.2.3 Insertion Sort

We now apply divide-and-conquer method to the "Insertion Sort" problem. $a[\,]$ is an unsorted array of length n. In short, the insertion sort puts $a[\,]$ at its proper place in the array $a[1 : j - 1]$ for $j = 2$ to n.

Algorithm-Insertion Sort is shown below.

Algorithm 2.5 Insertion sort

```
1.  Algorithm InSort (a, n)
2.  // a is an array of size n ≥ 1.
3.  // This algorithm sorts the array a in nondecreasing order
4.  {
5.      for k: = 2 to n do
6.      {
7.          key: = a[k];
8.              // insert a[j] in the sorted array a[1:j - 1]
9.          l: = k - 1;
10.             while l > 0 and a[l] > key do
11.             {
12.                 a[l + 1]: = a[l];
13.                    l: = l - 1;
14.             }
15.                 a[l + 1]: = key;
16.     }
17. }
```

Let us consider the time complexity of the insertion sort algorithm. The time complexity of any algorithm is the sum of the time complexities of each of the statements that contribute to the total time. Comments do not contribute anything to the total time complexity. In the said insertion sort algorithm the statement numbers, the cost associated with each statement and the number of times each statement repeats is listed below. Let us suppose that the cost associated with the statement numbered j is c_j. Let t_j be the number of times the loop is executed for the value of j. The time complexity of statement number j is $c_j t_j$.

Statement Number	Cost	Number of times repeated
5	c_5	n
7	c_7	$n-1$
9	c_9	$n-1$
10	c_{10}	$\sum_{j=2}^{n} t_j$
12	c_{12}	$\sum_{j=2}^{n} (t_j - 1)$
13	c_{13}	$\sum_{j=2}^{n} (t_j - 1)$
15	c_{15}	$n-1$

The time complexity $T(n)$ is

$$T(n) = c_5 n + c_7(n-1) + c_9(n-1) + c_{10} \sum_{j=2}^{n} t_j$$

$$+ c_{12} \sum_{j=2}^{n} (t_j - 1) + c_{13} \sum_{j=2}^{n} (t_j - 1) + c_{15}(n-1)$$

The best case occurs when the array a is already sorted. In this case for $j = 2, 3, \ldots, n$, $a[i] \le$ key in line 10 with l initially having the value $j-1$. Hence, $t_j = 1$ for $j = 2, 3, \ldots, n$. Therefore, the time complexity of the best case is

$$T(n) = c_5 n + c_7(n-1) + c_9(n-1) + c_{10}(n-1) + c_{15}(n-1)$$
$$= An + B \text{ where } A \text{ and } B \text{ are suitable constants}$$

Thus we see that the best case running time of the insertion algorithm is $O(n)$.

If the array $a[\,]$ is sorted in the decreasing order, then the worst case occurs. Then we have to compare each $a[j]$ with each element in the entire sorted subarray $a[1 : j-1]$. Hence $t_j = j$ for $j = 2, 3, \ldots, n$. Then

$$T(n) = c_5 n + c_7(n-1) + c_9(n-1) + c_{10}(n(n+1)/2 - 1)$$
$$+ c_{12}(n(n-1)/2) + c_{13}(n(n-1)/2) + c_{15}(n-1)$$
$$= Cn^2 + Dn + E \text{ for some suitable constants } C, D \text{ and } E.$$

Thus, the worst case time complexity of the insertion sort algorithm is $O(n^2)$.

2.2.4 Quick Sort

In the Quick Sort algorithm the merging part as is done in the Merge Sort is avoided. We select an index k, such that $1 \le k \le n$ and $a[i] \le a[j]$, for all i, $1 \le i \le k$ and for all j, $k + 1 \le j \le n$. The index k is called the partitioning index. The two arrays $a[1 : k]$ and $a[k+1 : n]$ can be sorted independently. We assume that all the n elements of the array $a[\,]$ are distinct and that the partitioning element $t = a[k]$ in the call of Split(a, k, p) has an equal probability of being the ith smallest element in $a[m, p-1]$. Regrouping of elements is necessary for this.

The regrouping of the elements of the given array $a[\,]$ is done as follows. We select one element, say $a[m]$ and rearrange the other elements, so that all the elements of $a[1:n]$ to the left of $a[m]$ are less than or equal to $a[m]$ and all elements appearing to the right of $a[m]$ are greater than or equal to $a[m]$. The two operations, namely "splitting" and "regrouping" are performed recursively.

In algorithm 2.6, the function Split partitions the elements of $a[k, p-1]$. Assuming $a[p] \leq a[k]$ and that $a[k]$ is the partitioning element. Initially Split $(a, 1, n+1)$ is invoked.

Algorithm 2.6 Partitioning of an array

```
Algorithm Split(a, k, p)
// The elements in a[k : p-1] are regrouped. This regrouping
// is done by selecting an element a[s] then regroup the
// elements so that all the elements that appear before a[s]
// are less than or equal to a[s] and all elements that appear
// after a[s] are greater than or equal to a[s].
{
    t: = a[k];
    i: = k;
    j: = p;
    repeat
    {
      repeat
        i: = i + 1;
      until (a[i]) ≥ t );
        repeat
          j: = j - 1;
        until (a[j] ≤ t);
      if (i < j) then swap(a[i], a[j]);
      }
    until (i ≥ j);
    a[k]: = a[j];
    a[j]: = t;
    return j;
}
```

We now give the QuickSort algorithm 2.7 that describes this process.

Algorithm 2.7 Quick sort

```
Algorithm QuickSort(i, j)
// This algorithm sorts the array elements a[p:q].
{
```

```
    if (i < j ) then // Small(P) is false. So, split.
    {
        loc := Split(a, i, j+1) ;
        // location of the partitioning element identified
        QuickSort(i, loc -1) ;
        QuickSort(loc+1, j) ;
    }
}
```

Example 2.2

We now trace the above algorithm for the sample data $a[\,] = (44, 20, 77, 15, 62, 53, 91, 35, 98, 84)$

The elements that are underlined are compared

$\underline{44}$	$\underline{23}$	77	15	62	53	91	35	98	84	
$\underline{44}$	23	$\underline{77}$	15	62	53	91	35	98	84	
$\underline{44}$	20	77	15	62	53	91	35	98	$\underline{84}$	
$\underline{44}$	20	77	15	62	53	91	35	$\underline{98}$	84	
$\underline{44}$	20	77	15	62	53	91	$\underline{35}$	98	84:	77 and 35 are interchanged
44	20	35	15	62	53	91	77	98	84	
$\underline{44}$	20	35	15	$\underline{62}$	53	91	77	98	84	
$\underline{44}$	20	35	15	62	53	$\underline{91}$	77	98	84	
$\underline{44}$	20	35	15	62	$\underline{53}$	91	77	98	84	
$\underline{44}$	20	35	15	$\underline{62}$	53	91	77	98	84	
$\underline{44}$	20	35	$\underline{15}$	62	53	91	77	98	84:	44 and 15 are interchanged
15	20	35	44	62	53	91	77	98	84	$i \geq j$.

Thus in this example, the function Split partitions the original set into two subsets $S_1 = \{15, 20, 35\}$ and $S_2 = \{62, 53, 91, 77, 98, 84\}$. The same method can be applied to each of these sub-sets until the original set of elements is totally sorted.

Whenever the function Split is invoked, it generates two sets S_1 and S_2 such that all the elements in S_1 are less than or equal to the elements in S_2. Hence, S_1 and S_2 can be sorted independently.

Time Complexity of the Quick Sort Algorithm

We count $C(n)$, the number of comparisons. We first obtain the worst case value $C_w(n)$ of $C(n)$. When a call to the function Split is made the maximum number of comparisons made is $p - k + 1$. If r denotes the number of elements in a call to Split at any level. then at the first level, Split(a, 1, $n + 1$) is referred only once and $r = n$. At each subsequent higher levels, the number gets decreased by one. We observe that $C_w(1) = 0$. Thus

$$C_w(n) = \Sigma r \text{ for } r = 2 \text{ to } n$$

$$= n(n+1)/2 - 1 = O(n^2)$$

Now consider the average time complexity $C_A(n)$. When Split(a, k, p) is called, two sub-arrays $a[k:j]$ and $a[m+1:p-1]$ are formed. The probability of this splitting is $1/(p-k)$. The number of comparisons, required on the first call of Split function is $(n+1)$. The recurrence relation for $C_A(n)$ is

$$C_A(n) = (n+1) + (1/n) \sum_{1 \le k \le n} [C(k-1) + C(n-k)] \tag{2.1}$$

Note that $C_A(0) = C_A(1) = 0$.

We solve this recurrence relation. Multiplying both sides of (2.1) first by n and then by $(n-1)$ and subtracting the later from the former, we get

$$\frac{C_A(n)}{n+1} = C_A \frac{n-1}{n} + \frac{2}{n+1} \tag{2.2}$$

This is a recurrence relation. Using the method of substitution, we get

$$\frac{C_A(n)}{n+1} = C_A \frac{n-2}{n-1} + \frac{2}{n} + \frac{2}{n+1}$$

$$= C_A \frac{n-3}{n-1} + \frac{2}{n-1} + \frac{2}{n} + \frac{2}{n+1}$$

$$= \frac{C_A(1)}{2} + 2 \sum \frac{1}{k}, \quad \text{taken over } 3 \le k \le n+1$$

$$= 2 \sum \frac{1}{k}, \quad \text{summation over } 3 \le k \le n+1$$

Therefore, $C_A(n) \le 2(n+1)$ [$\log_e (n+1) - \log_e 2$.

Hence, $C_A(n) = O(n \log_2 n)$.

2.3 HIGHEST- AND LOWEST-VALUED ELEMENTS IN AN ARRAY

We now apply the divide-and-conquer technique, for finding the maximum and minimum items in a given array of size n. As described earlier; we consider an arbitrary instance $P = (n, a[i:j])$ of the problem, where n denotes the number of elements in $a[i:j]$ and the largest and smallest elements in $a[i:j]$ are set to high and low respectively. Note that, if $n = 1$, then high = low = $a[1]$. We make one comparison if $n = 2$. Thus Small(P) is true if $n \le 2$. Otherwise, the array $a[i:j]$ is split recursively into smaller and smaller sub-arrays, until Small(P) is true. The recursive algorithm 2.8 is for finding the maximum and minimum of n elements.

Algorithm 2.8 A recursive algorithm for maximum and minimum

```
Algorithm RHighLow (p, q, high, low)
// a[1: n] is a given array. This algorithm returns
// the largest and smallest values in a[p : q], 1≤p≤q≤n.
{
    if (p = q) then // Sma (P) is true
    {
```

```
            high := a[p];
            low := a[p];
        }
        else
        if (p = q-1) then // Small(P) is again true
           {
              if (a[p] <a[q]) then
              {
                  high:=a[q] ;
                  low:=a[p];
              }
                  else
              {
                  high1:=a[p] ;
                  low1:=a[q] ;
              }
              }
                else // P not small
                mid := ⌊(p+q)⌋/2;
                RHighLow(p,mid,high,low) ;
                RHighLow(mid+1,j,high1,low1);
                //Merge all the sub-solutions
                if (high < high1) then high := high1 ;
                   if (low > low1) then low := low1;
           }
    }
```

Time Complexity of Recursive RHighLow Algorithm

An arbitrary instance of the problem is $P = (n, a[i], a[i + 1], ..., a[j])$. As already said, Small(P) is true, when $n \leq 2$. If the list is large, P has to be divided into smaller instances. For example, we may divide P into two instances

$$P_1 = ([n/2], a[1], ..., a[n/2])$$

and $$P_2 = (n - [n/2], a[n/2] + 1], ..., a[n])$$

After dividing P into two subproblems, we can invoke recursively, if necessary, the same divide-and-conquer algorithm.

Then we combine the solutions of the subprograms P_1 and P_2. If High(P) and Low(P) are the maximum and minimum of elements in P, then

$$\text{High }(P) = \text{larger (High }(P_1), \text{ High }(P_2))$$
$$\text{Min }(P) = \text{smaller (Low }(P_1), \text{ Low }(P_2))$$

The procedure is initially invoked by calling RHighLow $(1, n, x, y)$.

$T(n)$, the number of comparisons made in RHighLow is given by the recursive relation:

$$T(n) = \begin{cases} T[n/2] + T[n/2] + 2 & \text{if } n > 2 \\ 1 & \text{if } n = 2 \\ 0 & \text{if } n = 1 \end{cases}$$

When n is a power of 2, $n = 2^k$ for some positive integer k, then one can solve the above recurrence and obtain

$$T(n) = 2^{k-1}T(2) + \sum_{1}^{k-1} 2^i$$

$$= 2^{k-1} + 2^k - 2$$

$$= 3n/2 - 2$$

Note that $3n/2 - 2$ is the best average and worst case number of comparisons when n is a power of 2.

2.4 STRASSEN'S MATRIX MULTIPLICATION

Let X and Y be the two matrices each of order $n \times n$. We want of output the product matrix $Z = X * Y$. The important steps in the algorithm of obtaining the product matrix in the direct and commonly used method are as follows.

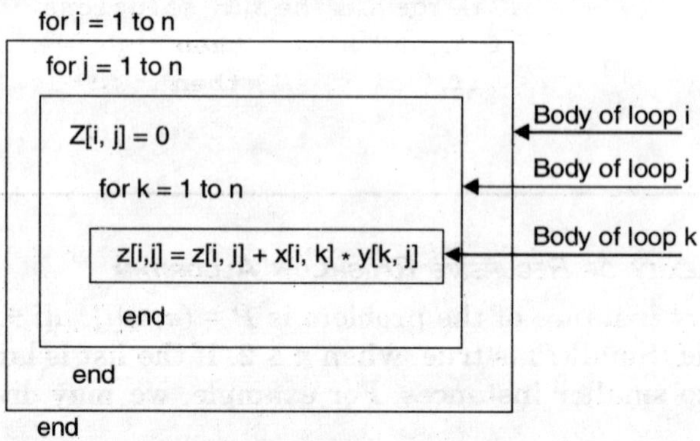

There is only one step in the body of the inner loop (loop of k) and it takes 4 steps to access it. This step involves 4 accessions to the variables and 2 basic arithmetic operations. Thus as seen above, the number of steps involved in the inner loop is:

$$4 * 4 + 2 = 18$$

Thus, the time of execution of the innermost loop is

$$2 + n(18 + 3) = 21n + 2$$

For loop j there are 4 accessions to variables and 1 assignment (total 5). So, the number of instructions in the body of loop j is $21n + 7$.

The body of the outermost loop contains

$$2 + n(21n + 7 + 3)$$

$$= 21n^2 + 10n + 2$$

The final and total time complexity of all the three loops is:

$$2 + n(21n^2 + 10n + 2)$$
$$= 21n^3 + 10n^2 + 2n + 2$$

Leaving all the lower order terms and numerical coefficient, we say that the time complexity of this algorithm for matrix multiplication is of order $\Theta(n^3)$. We say that this is of cubic order. The alternate methods that use lesser order of times of execution are devised. Below we discuss one such method.

The method is applying divide-and-conquer technique. For computational facility, we assume that $n = 2^k$ for some positive integer k. If necessary we append necessary number of rows or columns to both X and Y all of zero elements. We split each matrix into four matrices, each of size $n/2 \times n/2$. Then the product $X * Y$ can be computed recursively, by using the above formula for the product of 2×2 matrices. If $X * Y$ is

$$\begin{bmatrix} X_{11} & X_{12} \\ X_{21} & X_{22} \end{bmatrix} * \begin{bmatrix} Y_{11} & Y_{12} \\ Y_{21} & Y_{22} \end{bmatrix} = \begin{bmatrix} Z_{11} & Z_{12} \\ Z_{21} & Z_{22} \end{bmatrix} \tag{2.3}$$

where

$$\left.\begin{aligned} Z_{11} &= X_{11}Y_{11} + X_{12}Y_{21} \\ Z_{12} &= X_{11}Y_{12} + X_{12}Y_{22} \\ Z_{21} &= X_{21}Y_{11} + X_{22}Y_{21} \\ Z_{22} &= X_{21}Y_{12} + X_{22}Y_{22} \end{aligned}\right\} \tag{2.4}$$

As discussed in the earlier examples, the problem is split recursively into smaller subproblems (less dimension), until Small(P) is true. We suppose that when $n = 2(k = 1)$, Small(P) is true and product matrix can be obtained directly, using formulae (2.1) and (2.2).

If $n > 2$, and $n = 2^k$, Small(P) is false and P is split recursively into subproblems. At each stage we have elements which are $n/2 \times n/2$ matrices. The fact that n is a power of 2 allows us to compute these matrix products recursively using the same procedure that was used for $n \times n$ matrices. We apply this algorithm continuously again and again until n becomes suitably small ($n = 2$) so that the product can be computed directly, using formulae (2.1) and (2.2).

However, even in the above method, we observe that there are eight multiplications and four additions each being a matrix of order $n/2 \times n/2$. Two $n/2 \times n/2$ matrices can be added in time bn^2 for some constant b. The computing time $T(n)$ of the above method is given by the recurrence relation

$$T(n) = \begin{cases} a & \text{for } n \le 2 \\ 8T(n/2) + bn^2 & \text{for } n > 2,\ a \text{ and } b \text{ are constants} \end{cases}$$

Solving this recurrence relation, we obtain $T(n) = O(n^3)$ (cubic order). Thus, we notice that no reduction in the computing time is achieved.

Volker Strassen gave a reformulation of the formula (2.2) for computing Z_{ij}, that involves less number of multiplications and may be more number of additions. Increase in the number of matrix additions is not a matter to worry as matrix multiplications are more expensive than matrix additions, ($O(n^3)$ versus $O(n^2)$).

For multiplication of two 2 × 2 matrices his method uses only 7 multiplications and 18 additions or subtractions. Now we explain this method. We first compute the seven $n/2 \times n/2$ matrices P, Q, R, S, T, U and V as follows:

$$P = (X_{11} + X_{22})\,(Y_{11} + Y_{22})$$
$$Q = (X_{21} + X_{22})\,Y_{11}$$
$$R = X_{11}\,(Y_{12} - Y_{22})$$
$$S = X_{22}\,(Y_{21} - Y_{11})$$
$$T = (X_{11} + X_{12})\,Y_{22}$$
$$U = (X_{21} - X_{11})\,(Y_{11} + Y_{12})$$
$$V = (X_{12} - X_{22})\,(Y_{21} + Y_{22})$$

Finally, we compute the elements of $Z = X * Y$ matrix as

$$Z_{11} = P + S - T + V$$
$$Z_{12} = R + T$$
$$Z_{21} = Q + S$$
$$Z_{22} = P + R - Q + U$$

TIME COMPLEXITY OF STRASSEN'S MATRIX MULTIPLICATION METHOD

The computing time $T(n)$ of this method is given by the recurrence relation

$$T(n) = \begin{cases} a & \text{if } n \le 2 \\ 7T(n/2) + bn^2 & \text{if } n > 2 \end{cases}$$

where a and b are constants. We now solve this recurrence as

$$T(n) = bn^2[1 + 7/4 + (7/4)^2 + \cdots + (7/4)^{k-1}] + 7^k T(1)$$
$$\le cn^2 (7/4)^{\log n} + 7^{\log n}, \ c \text{ being a constant.}$$
$$= cn^{\log 4 + \log 4} + n^{\log 7} = O(n^{\log 7}) \approx O(n^{2.81})$$

(all logarithms are to the base 2).

We observe that the time complexity of Strassen's method is only $O(n^{2.81})$ where as the time complexity for the other two methods is $O(n^3)$. This is a great saving, particularly when n is sufficiently large.

2.5 THE CONVEX-HULL PROBLEM

Let X be a set of points in a plane. The convex hull of X, denoted by CH(X), is the smallest convex polygon P for which each point of X is either a vertex of

P (such points are also called the extreme points) or is inside P. It is evident that all the extreme points of P are also points of X. There may or may not be points inside P. Now the problem is to compute the convex hull CH(X) of a set X of n points (i.e. to find all vertices of P). Two algorithms are available for solving this problem. The first algorithm, called the Gram's scan algorithm runs in $O(n \log n)$ time. The second one, called the Jarvis's march runs in $O(n*h)$, where h is the number of vertices of the convex hull.

2.5.1 Graham's Scan Algorithm

In solving the convex-hull problem, the Graham's scan algorithm maintains a stack S of points of X. Initially, each point of X is pushed into the stack. Then, the points that are not vertices of CH(X) are popped from the stack. When the algorithm terminates, the stack S contains exactly the vertices of CH(X), in counterclockwise order of their appearance on the boundary.

In the following algorithm, the input is the set X of n points in the two-dimensional plane, where $|X| \geq 3$. The function TOP(X) is called to return the top point of the stack S and the function NEXT-To-TOP is called to return the point one entry below the top of stack S. In both the calls the stack S is not changed. Finally, the stack S contains exactly the vertices of CH(X) from bottom to top, in the counterclockwise order.

Algorithm 2.9 Graham's scan algorithm—Convex hull problem

```
Algorithm GrahamScan (X)
// p0 is a point with minimum y-coordinate.
// { p1, ..., pm} is the set of the remaining points in X, sorted by
// polar angle in counterclockwise order around p0
// if there are more than one such points, remove all but the one
// that is farthest from p0
PUSH (p0, S); PUSH (p1, S); PUSH (p2, S)
for i = 3 to m do
do while the angle formed by points NEXT-To-TOP(S), TOP(S), and
pi makes a nonleft turn
   do POP (S)
 PUSH (pi, S)
return S
```

2.6 CONTROL ABSTRACTION OF DIVIDE-AND-CONQUER METHOD

Having considered few examples, wherein the divide-and-conquer technique is applied and got some idea of this technique, we now discuss the control abstraction of this technique. As explained in the beginning of this chapter, Small(P) is a boolean function. Small(P) is true means that the input size is small enough so that the problem P can be solved directly without any further splitting. Then, in the following algorithm, the function S is invoked. On the other hand, if Small(P) is false, the problem P is divided recursively, into smaller and smaller sub-problems and are solved again recursively using D and C as given below.

Algorithm 2.10 Control abstraction of divide-and-conquer

```
AlgorithmDandC(P)
//Gives the control abstraction of the divide-and-conquer method.
{
  if Small(P) then return S(P);
  else
  {
     divide P into smaller instances P1, P2, …, Pk ;
     Apply DandC to each of the smaller instances ;
     return Combine(DandC(P1), D and(P2), …, D and C(Pk));
  }
}
}
```

In general, the time complexity of a divide-and-conquer algorithm is given by a recurrence relation of the form

$$T(n) = \begin{cases} T(1) & \text{for } n = 2 \\ aT(n/b) + f(n) & \text{for } n > 1, a, b \text{ are known constants} \end{cases}$$

The function $f(n)$ in the above recurrence refers to the time taken for splitting and combining sub-solutions. $T(1)$ is known. For computational simplicity of solving recurrence relations we assume that n is equal to a power of b ($n = b^k$).

There is a number of methods for solving a recurrence relation depending on their types. One such method, that is used widely, is the substitution method. We solve below a few recurrence relations.

2.7 CLOSEST-PAIR OF ELEMENTS

Given a set Q of points (x_k, y_k) for $k \geq 2$, we design an algorithm that is fairly fast and gives correct output, for finding the closest-pair of points. Here the term "closest" is used with reference to the distance between two points. To avoid the trivial case, we suppose that no two points coincide. We apply the divide and conquer method in the design of this algorithm.

We use a recursive algorithm. Each recursion, takes an input $P \subseteq Q$ and arrays X and Y. The arrays X and Y contain all points of the input subset P. Points in X are sorted on their x-coordinates and those of Y on their

y-coordinates. We first check whether $|P| \leq 3$. If this is true, all ${}^{|P|}C_2$ pairs of points are verified for the "closest-pair" test. On the other hand, if $|P| > 3$, apply the divide and conquer algorithm stepwise as described below.

2.7.1 Algorithm of Closest-pair of Points

Step 1: Find a line l that bisects the points of the set P into two sets P_L and P_R such that $|P_L| = \text{Ceil}(|P/2|)$ and $|P_R| = \text{floor}(|P/2|)$. The set P_L contains all points of P that are either on the line l or to the left of it. Similarly, all points in P_R are either on the line l or to the right of it.

Step 2: Divide the array X into two arrays X_L and X_R each of which contains the points of P_L and P_R respectively. Sort the elements of X_L and X_R on their x-coordinates in the increasing order. Divide the array Y into two arrays Y_L and Y_R each of which contains the points of P_L and P_R respectively. Sort the elements of Y_L and Y_R on their y-coordinates in increasing order.

Step 3: Having divided P into P_L and P_R, make two recursive calls that correspond to the actions of finding the closest-pair of points in P_L and closest-pair of points in P_R. Let the closest-pair distances returned by P_L be δ_L and the one returned by P_R be δ_R. Take $\delta = \min(\delta_L, \delta_R)$. We observe that the inputs for the first call are P_L, X_L, and Y_L and the inputs for the second call are P_R, X_R, and Y_R. Notice that δ corresponds to a pair of points, one point in P_L and the other point in P_R. We now check whether there is a pair of points whose distance is less than δ. If such a pair exists, they must be within a distance of δ on both sides of the line l. This is shown in Figure 2.1

Step 4: Consider a vertical strip of width 2δ centred at line l. Then the required points must be within this vertical strip. One point is to the left of l and another point is to the right of l.

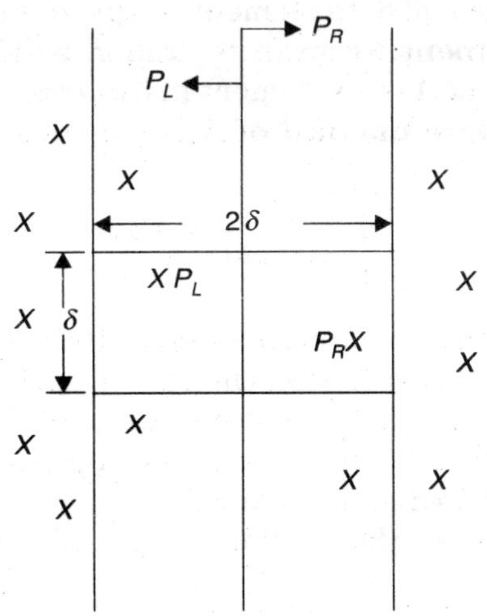

Figure 2.1 The "divide" part of the divide-and-conquer algorithm for the problem.

Step 5: Find the closest-pair of points by implementing the following steps:

Step 5.1: Let Y' be the array whose elements form a subset of the array elements of Y. Sort the elements of Y' based on their y-coordinates.

Step 5.2: For each point p in Y' mark (find) all points in Y' that are within a distance of δ from p. Compute the distance from p to each such point, paying special attention to the closest-pair distance δ'

Step.5.3: If $\delta' < \delta$, (the vertical strip contains the closest-pair than that was obtained by a recursive call.)
then
return the pair and the distance δ'
else
return the closest-pair and the distance δ as found by the recursive calls.

Time Complexity of this Algorithm

As seen in the earlier cases, the recurrence relation of $T(n)$ can be derived as

$$T(n) = 2\,T(n/2) + O(n) \quad \text{if } n > 3$$
$$= O(1) \quad \text{if } n \geq 3$$

whose solution is

$$T(n) = O(n \log_2 n)$$

2.8 FINDING "MEDIAN" OF A SET OF DATA ITEMS

All the intricate procedural aspects and their implementations employed in the "divide-and-conquer" technique are well understood with the following problem called "**Median finding**". It is hoped that the reasoning used in this problem will help the user in applying the divide-and-conquer method more purposely.

An array A, of size n of data elements is given. First sort the elements. By the **rank of an element**, we mean its position in the sorted array. **Median** of the array elements of A is an element in the array whose rank is $\lfloor n/2 \rfloor$.

By an **approximate median of** A, we mean an element x in A such that

$$\frac{n}{4} \leq \text{rank}(x) \leq \frac{3n}{4}$$

Let x be the approximate median in A. Split all the elements of A into two parts, namely A_L and A_R, such that the part A_L contains all elements in A which are less than x and the other part A_R contains those elements in A which are greater than x. We first obtain the approximate median of A and from it the actual median is obtained.

If rank(x) $< n/2$, then the median lies in A_R and if $x > n/2$, then the median lies in A_L.

Note that the smaller part is at least of size $n/4$ and the bigger part is at least of size $3n/4$.

Suppose the approximate median is found in linear time. Then the time $T(n)$ required to find the actual median is obtained from the recurrence relation

$$T(n) = n + T(3n/4)$$

The first term on the right-hand side is the time for finding the approximate median. It can be seen that $T(n)$ is $O(n)$. Now, suppose the actual median of A is in A_R. We can find the rank of the actual median that is in A_R. The rank of any element in A_R depends on the size $|A_L|$. Notice that $|A_L| < \lfloor n/2 \rfloor$. If the rank of the actual median is r, then

$$r + 1 + |A_L| = \lfloor n/2 \rfloor$$

This gives that $r = \lfloor n/2 \rfloor - |A_L| - 1$.

The rank of the median in A is $n/2 - |A_L| - 1$.

By finding the element in A_R of rank $n/2 - |A_L| - 1$, we can find the rank of the median. Such an element need not be at the middle of the array A. We now look at the general problem of finding the element in A_R of rank equal to $n/2 - |A_L| - 1$.

Split the elements of the array A into $n/5$ groups. The first group $G(1)$ of size 5, the second group $G(2)$ of size 5 and so on with the last group $G(n/5)$. Let $m_1, m_2, \ldots, m_{n/s}$ be the medians of each of these groups respectively. Now, we have an array of medians. Notice that the size of the new array is less than n (size reduction). The elements in each group are arranged as shown. One of the groups, if not the last group, may have less than 5 elements. All groups will contain 5 elements if n is divisible by 5. Sort the elements in increasing order. Consider the middle elements: $a_{13}, a_{23}, \ldots, a_{n/5\,3}$. Sort all these middle elements in increasing order. The middlemost element in A can be in any group. To facilitate the easy working, we take it as a_{13}.

$$\begin{array}{cccccccc}
a_{15} & a_{25} & \cdots & \cdots & & \cdots & \cdots & a_{n/s\,5} \\
v & v & & & & & & \\
a_{14} & a_{24} & \cdots & \cdots & \cdots & & & a_{n/s\,4} \\
v & v & & & & & & \\
\mathbf{a_{13}} < & \mathbf{a_{23}} < & \mathbf{a_{33}} < & \cdots & \cdots \mid \; \mid & \cdots & \cdots & \mathbf{a_{n/5\,3}} \\
v & v & & & & \cdots & \cdots & v \\
a_{12} & a_{22} & \cdots & \cdots & \cdots & \cdots & \cdots & a_{n/5\,2} \\
v & v & \cdots & \cdots & \cdots & \cdots & \cdots & v \\
a_{11} & a_{21} & \cdots & \cdots & \cdots & \cdots & \cdots & a_{n/5\,1} \\
G(1) & G(2) & \cdots & \cdots & \cdots & \cdots & \cdots & G(n/5)
\end{array}$$

The median of $\{a_{13}, a_{23}, a_{33}, a_{43}, \ldots, a_{n/5\ 3}\}$ is the approximate median of the array A. Let it be $a_{n/10\ 3}$. That is

$$a_{13} \quad a_{23} \quad \ldots \quad | \; | \quad \ldots \quad a_{n/5\ 3}$$

$$\rightarrow a_{n/5\ 3}$$

Pick up the element $a_{n/5\ 3}$. The elements which are larger than $a_{n/5\ 3}$ are located at the upper left-hand corner and those that are less than $a_{n/5\ 3}$ are at the bottom right-hand corner.

We are now equipped with all that is necessary to design the algorithm for finding the median of the array A.

FindElement(r, A)
/* Given an array A of dimension n and a positive integer r */
/* this algorithm finds the element in A whose rank is r */
Partition A into groups each of size 5
/* One of the groups, if not the last has less than 5 elements */
Find the median of each of these (n/5) groups and put them in array B.
FindElement (n/10, B)

Now, put this in an appropriate loop. This algorithm is used to find the approximate median x, of A using a partitioning technique. Partition A into two sub-arrays A_L and A_R such that the elements in A_L are all less than x, and the elements in A_R are all greater than x.

Depending upon the rank of x, we recourse either on A_L or on A_R. Then three cases arise. The description of the three cases and the actions taken in each case are:

Case-1: **If** the size of the sub-array, $|A_L| = r - 1$, **then** return n.
Case-2: **If** $|A_L| > r - 1$, **then FindElement** (r, A_L).
Case-3: **Else** ($|A_L| < r - 1$), **FindElement** ($r - |A_L| - 1, A_R$)

We claimed that $T(n)$ for this algorithm is $O(n)$, a linear order. The justification of our claim is done as follows. We observe that the time order for partitioning is $T(n/5)$ and the one for finding the median of a sub-group is $5n$. Hence, we take the recurrence relation for $T(n)$ as

$$T(n) = T(n/10) + 5n$$

The maximum size in A_L or in A_R is $3n/4$. Then it is easy to see that $T(n)$ is composed of the terms for the time complexity for finding the median of n/5 sub-groups (c_1n), the one for finding the actual median ($T(n/5)$) and the one for partitioning (n) and finally the time complexity for the recurrence ($T(3n/5)$). We thus have

$$T(n) = c_1n + T(n/5) + n + T(3n/4)$$

The recurrence is done at two levels, the size in the first level being and that for the second level being.

Total size is

$$\frac{n}{5} + \frac{3n}{4} = \frac{19n}{20} < n$$

We are successful and we claim that we end up finally at something that is linear in n. Let us prove this claim by showing that $T(n) \leq cn$, for some constant c. From the above inequality for $T(n)$, we write

$$T(n) \leq (c_1 + 1)n + c(n/5)\, c(3n/4)$$
$$\leq cn$$

If this is established, the task is done. For this to be true we must have

$$(c_1 + 1) \leq c - (n/5) - (3n/4)$$
$$c_1 + 1 \leq c/20 \text{ or } c \geq 20(c_1 + 1)$$

We can always choose c and c_1 in such a way so that

$$T(n) \leq 20\,(c_1 + 1).$$

Thus $T(n)$ is linear in n.

Example 2.3

Consider the recurrence relation

$$T(1) = 2 \text{ and}$$
$$T(n) = 2T(n/2) + n \text{ if } n > 1$$

We solve this recurrence relation using the substitution method.

$$\begin{aligned} T(n) &= 2T(n/2) + n \\ &= 4T(n/4) + 2n \\ &= 8T(n/8) + 3n \\ &= \ldots\ \ldots\ \ldots \\ &= \ldots\ \ldots\ \ldots \\ &= 2^i\, T(n/2^i) + i\, n \text{ for any } \log_2 n \geq i \geq 1 \text{ where } n = 2^i. \end{aligned}$$

By taking $i = \log_2 n$, we have

$$T(n) = 2^{\log_2 n}\, T(n/2^{\log_2 n}) + n \log_2 n$$

Therefore,

$$T(n) = nT(1) + n \log_2 n^n = 2n + n \log_2 n$$

Example 2.4

Consider the recurrence relation, when n is a power of 2.

$$T(n) = \begin{cases} T(1) & \text{if } n = 1 \\ T(n/2) + a & \text{if } n > 1 \end{cases}$$

Following the aforesaid procedure of substitution and taking $a = 1$, $b = 2$, $f(n) = a$, we derive $T(n)$ as

$$T(n) = {}_n\log_b c\ [a + \Theta(\log n)] = \Theta(\log n)$$

If $a = 2$, $b = 2$ and $f(n) = cn$, then we obtain $c = \Theta(1)$ and $T(n) = \Theta(\log n)$.

Example 2.5

Consider the recurrence relation: $T(n) = 6T(n/2) + 18n^2$, $n \geq 2$ and n being a power of 2. Here, we have $a = 6$, $b = 2$ and $f(n) = 18n^2$. Therefore, $\log_b a = \log_2 6 = c$, say $h(n) = 18n^2/n^c = 18n^{2-c} = O(n^r)$, where $r = 2 - \log_2 6 < 0$. So $u(n) = O(1)$ and $T(n) = n^c[T(1) + O(1)] = \Theta(n^c)$, assuming $T(1)$ to be a constant.

Example 2.6

Consider the recurrence: $T(n) = 25T(n/5) + 4n^6$, for $n \geq 5$, and n is a power of 5.
Here, we have $a = 25$, $b = 5$ and $f(n) = 4n^6$.

$$\log_a b = 2,\ h(n) = 4n^6/n^2 = 4n^4 = \Omega(n^4)$$

From the above table, we observe that

$$T(n) = n^2[T(1) + \Theta(n^4)] = \Theta(n^6),$$

as $T(1)$ can be taken as a constant.

Exercises

2.1 Solve the recurrence relation
$T(n) = T(1)$, if $n = 0$ and $T(n) = aT(n/b) + f(n)$ if $n \geq 1$, for the following choices of a, b, and $f(n)$, c being a constant.
(a) $a = 1$, $b = 2$, $f(n) = cn$
(b) $a = 5$, $b = 4$, $f(n) = cn^2$
(c) $a = 28$, $b = 3$, $f(n) = cn^3$
(d) $a = 2$, $b = 2$ and $f(n) = cn$
(e) $T(n) = T(1)$ for $n = 1$ and $T(n) = T(n/2) + c$ for $n > 1$.
(f) $T(n) = 9T(n/3) + an^6$, $n \geq 3$ and a power of 3.

2.2 Solve the following relations by substitution method:

(a) $$T(n) = \begin{cases} 1 & \text{if } n \leq 4 \\ T(\sqrt{n}) + c & \text{if } n > 4 \end{cases}$$

(b) $$T(n) = \begin{cases} 1 & \text{if } n \leq 4 \\ 2T(\sqrt{n}) + \log n & \text{if } n > 4 \end{cases}$$

(c) $$T(n) = \begin{cases} 1 & \text{if } n \leq 4 \\ 2T(\sqrt{n}) + \log n/\log\log n & \text{if } n > 4 \end{cases}$$

QUESTIONS

1. Explain "Divide and Conquer" method and write an algorithm for its control abstraction and also discuss its time complexity.
2. Write an algorithm for finding the maximum and minimum of a sequence of given numbers and analyse its average and worst time complexities.
3. Write an algorithm for the binary search and discuss its time complexities of worst, average and best cases.
4. Analyse the average case time complexity of Quick sort algorithm.
5. Write an algorithm for the insertion sort and discuss its average and worst time complexities.
6. Write the "Merge sort" algorithm and suggest refinements to make it in-place.
7. Give the selection sort algorithm and discuss its space and time complexities.
8. Explain Strassen's matrix multiplication method and discuss its advantages over the conventional method.
9. Discuss the Convex-Hull problem and write the Graham scan's algorithm.

CHAPTER 3

The Greedy Method

3.1 INTRODUCTION

Greedy method is a powerful technique used in the design of algorithms. Almost all problems that come under this category have n inputs. A subset of the given set of inputs that satisfies some given constraints is to be obtained. Such a subset is called a **feasible solution**. Now the problem is to find a feasible solution that maximizes or minimizes a given objective function. Such a function is called an **optimal solution**.

The greedy method works in stages. Suppose we have a partially constructed optimal solution. The next input from set of remaining inputs is selected and is tested whether it contributes to the optimality of the solution of the problem. Otherwise this input is added to the above said partially constructed optimal solution. This input is rejected and is not added to the partially constructed optimal solution. For this, we need a selection procedure. This selection procedure determines the order of the inputs. Initially the partially constructed optimal solution is an empty set. At the first stage it considers the first input in the list and in the second stage it considers the second input in the list and this process continues till all the inputs in the list are exhausted. Note that if inclusion of an input gives a feasible solution, then only it is added to the partial optimal solution set otherwise not. This selection procedure is called the **subset paradigm of the greedy technique**. The control abstraction of the greedy method is given in algorithm 3.1.

Algorithm 3.1 Control abstraction of Greedy method

```
AlgorithmGreedy(a, n)
// This algorithm gives the control abstraction of the Greedy
method.
// The input array is a[1 : n].
{
  answerset := Ø;
for j: = 1 to n do
{
  x: = Select(a);
    if Feasible (solution, x) then
    answerset: = Union( answerset, x );
}
return answerset;
}
```

In the above algorithm, the function **Select** selects and removes an input from the array $a[\,]$ and assigns it to x. **Feasible** is a boolean function that determines whether x can be included into the partially constructed optimal solution vector. The function **Union** adjoins x to the aforesaid solution vector.

In this chapter, we apply greedy method to (1) the Knapsack Problem, (2) Tree Vertex Splitting Problem, (3) Job Sequencing Problem, (4) Minimum-Cost Spanning Trees. (5) The Optimal Storage Problem and (6) Single Source Shortest Paths.

3.2 THE KNAPSACK PROBLEM

We are given n objects and a knapsack (bag) having a maximum capacity of m. Object i has weight w_i. Entire object i or a fraction x_i of the object i can be selected and placed in the knapsack. If a fraction x_i, $0 \le x_i \le 1$, of object i is placed in the knapsack, then a profit of $p_i\, x_i$ is earned. The problem is to decide what fraction of each object is to be put in the knapsack so as to earn a maximum total profit. Note that the maximum total weight of all chosen objects cannot exceed m.

The mathematical formulation of the knapsack problem is:

$$\text{Maximize } \Sigma p_i\, x_i$$

$$\text{subject to } \Sigma w_i\, x_i \le m$$

$$\text{and for } 0 \le x_i \le 1, \qquad \text{for } 1 \le i \le n.$$

An optimal feasible solution is any set $(x_1, x_2, \ldots, x_n)$ satisfying the above constraints.

Here x_i represents the fraction of the ith-item selected and put in the knapsack. We use greedy algorithm and pick one item by the other. The delicate question that arises in this context is the selection process of the next item. We sort the items either in the decreasing order of profits or in the increasing order of weights (otherwise called sizes). Assume that y_t units of item t are picked to get

$$\sum_1^n y_t\, w_t = m$$

with profit

$$\sum_1^n y_t\, p_t$$

The units of items chosen that make the knapsack full are:

$$y_1 \quad y_2 \quad \cdots \quad y_i \quad \cdots \quad y_j \quad \cdots \quad y_n$$

Now we use the exchange trick. Let ε units of unit "i" are discarded and let ε' units of item "j" are added to the knapsack hoping still to make the knapsack full. Then, we must have

$$y_1w_1 + y_2w_2 + \cdots + (y_i - \varepsilon)\, w_i + \cdots + (\, y_j + \varepsilon'\,)_j w_j + \cdots + y_nw_n = m$$

This implies that

$$\sum_1^n y_t w_t + w_i\, \varepsilon' - w_j\, \varepsilon = m$$

This is possible only if

$$w_i\, \varepsilon' - w_i\, \varepsilon = 0$$

That is

$$\varepsilon/\varepsilon' = w_i/w_j$$

The new profit is

$$p_1y_1 + p_2y_2 + \cdots + p_i(y_i - \varepsilon) + \cdots + p_j(y_j + \varepsilon') + \cdots + p_ny_n$$

$$= \text{Earlier profit} + (p_j\varepsilon' - p_i\varepsilon)$$

There will be an increase in the profit only if

$$p_j\varepsilon' > p_i\varepsilon$$

That is

$$p_j/p_i > \varepsilon/\varepsilon' = w_j/w_i$$

or

$$p_j/w_j > p_i/w_i.$$

In this context, we define score (item) as follows:

Score (item) = profit/weight

Now sort the items in decreasing order of scores. Then the knapsack is filled full to its capacity m.

The whole of one item after another is picked. A stage will come when by picking the whole of the next item, the total capacity is greater than m. So, a fraction of this item alone must be picked. The rest of the items are discarded (not picked up). We have shown this below.

The items selected in the decreasing order of score are:

x_1	x_2 ...	x_{j-2}	x_{j-1}	e_x	x_{j+1}	x_{j+2}	x_n
1	1	1	1	δ	0	0	0

where $0 < \delta < 1$.

It can be shown that in this way, one can obtain an optimal solution to the knapsack problem. In this context, we have the following theorem.

Theorem: If $p_1/w_1 \geq p_2/w_2 \geq \cdots \geq p_n/w_n$, then there is an optimal solution to the knapsack problem.

The next problem we take up is in the field of information transmission and is given by Huffman. This is solved by the greedy method. The problem is given as follows:

We are given n symbols which are otherwise called words: $x_1, x_2, \ldots, x_n$ with frequencies $f_1, f_2, \ldots, f_n$. These words are not transmitted as they are. A binary code is associated with each symbol and thus n binary codes are transmitted. This is done to maintain some sort of secrecy in the transmission. Thus the given set-up is:

Symbol	:	x_1	x_2	x_n
Frequency	:	f_1	f_2	f_n
Binary code	:	c_1	c_2	c_n

Consider the following example:

Symbol	:	N	P	C
Frequency	:	5	50	5
Code	:	11	0	10

Suppose we transmit a binary string. Scan the binary bits one by one from left to right. If it is 0, then associate P with it. If the bit scanned is 1, then scan the next bit to decide whether these two bits (10 or 11) correspond either to C or N. With this understanding we transmit the binary string 110010001011 to mean the string "NPPCPPCN".

A binary code C_i is said to be a prefix-free code, if for all other binary codes C_j, C_i is not a prefix of any other code. There is a close connection between "Binary trees" and "Prefix-free" binary codes. In any binary tree, the code words are at the leaves. From a node, we branch to the left on "0" and to the right on "1".

Now, our problem is:

$$\text{maximise} \sum_{i=1}^{n} h_i f_i$$

where h_i is the height of the symbol I in the binary tree.

The symbols with large frequencies are placed close to the root and they are associated with smaller length codes.

Thus, our problem turns out as follows:

Input: n frequencies
Output: Prefix-free binary codes, or otherwise a binary tree with n or more leaves and with all codes of symbols (words) at their leaves.

Before coming to the "exchange trick", we do the following:

Sort the symbols in increasing order of heights and in decreasing order of frequencies. Then, there are a number of binary trees with n or more leaves. What all we have to do is to select the correct binary tree.

Now consider a binary tree with a leaf at a height h_1 and a leaf at the bottommost of the binary tree at a height h_2 ($h_1 < h_2$). We then associate f_2 with h_1 and f_1 with h_2 such that

$$f_2h_1 + h_2f_1 < f_1h_1 + f_2h_2$$

Consider the actual binary tree having two sub-trees, T_i having height h and T_j having height h' and suppose that $h < h'$. Now apply the exchange trick on exchanging the two sub-trees T_i and T_j hoping to obtain a new prefix-free code with less cost. This configuration is shown below:

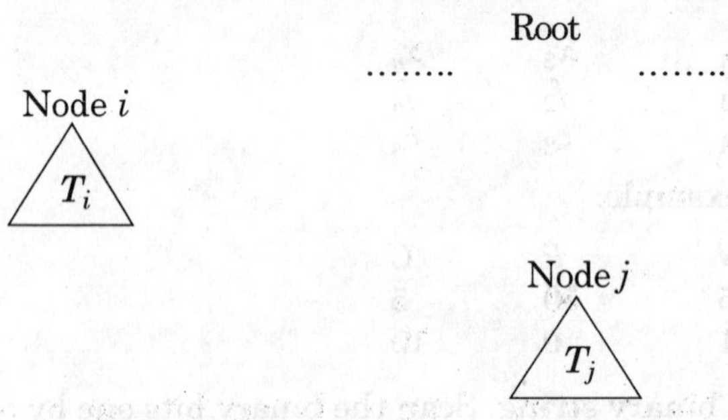

$$\text{Old cost} = \sum_{e\,\varepsilon\,T_i} f_e(h + l_e) + \sum_{e\,\varepsilon\,T_j} f_e(h' + l_e')$$

where l_e is the length of the element e in T_i from node I and so is the second term in the second summation in T_j.

$$\text{New cost} = \sum_{e\,\varepsilon\,T_i} f_e(h' + l_e) + \sum_{e\,\varepsilon\,T_j} f_e(h + l_e')$$

Difference in costs is

$$= (h' - h)\left(\sum_{e\,\varepsilon\,T_j} f_e + \sum_{e\,\varepsilon\,T_i} f_e\right)$$

The factor $(h' - h)$ is positive. Hence, there is a cost reduction if the second term in the above step is negative. That is,

$$\sum_{e\,\varepsilon\,T_j} f_e > \sum_{e\,\varepsilon\,T_i} f_e$$

If the above inequality is satisfied, then there will be a cost reduction by exchanging the two sub-trees T_i and T_j. Consider the frequencies

$$f_1 \quad f_2 \quad f_{n-2} \quad f_{n-1} \quad f_n$$

We combine the last two frequencies f_{n-1} and f_n to get f'_{n-1}. That is $f'_{n-1} = f_{n-1} + f_n$.

We chose two smallest among $f_1, f_2, \ldots, f_{n-2}, f'_{n-1}$ and combine them. The generic step in the algorithm is given sub-trees and their frequencies as

Sub-Tree	:	T_1	T_2	T_k
Frequency	:	f'_1	f'_2	f'_k

where

f'_i = Sum of frequencies of elements/symbols/words in the sub-tree T_i

Now, two sub-trees T_i and T_j are selected, where one of I, j is less than or equal to k and the other is less than k. These two sub-trees are merged to obtain reduction. When the algorithm terminates, we get the required binary sub-tree at the bottom of the original binary tree.

We now check that the algorithm works well and is correct. We can use the induction method to prove this fact.

Basis: At the base, we have at least n leaves and this must happen even in the optimum case. So, there is a base for our claim.

Inductive Step: Let us suppose that the optimum exists at the kth level.

Induction: There exists an optimal tree. Say T with $T_1, T_2 \ldots, T_k$.

Consider the following set up:

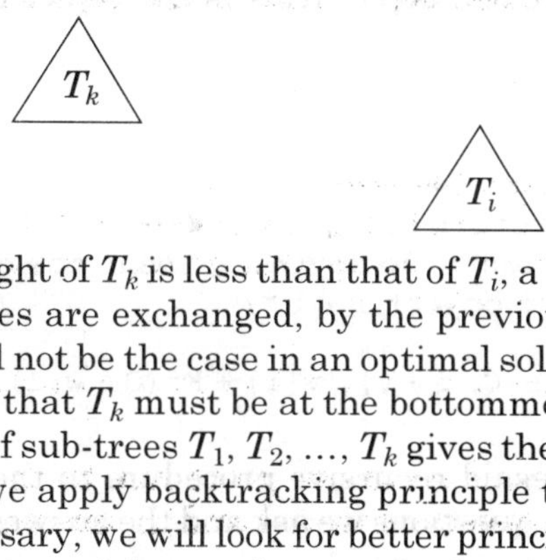

Suppose the height of T_k is less than that of T_i, a bottommost sub-tree. When these two trees are exchanged, by the previous argument the cost reduces. This should not be the case in an optimal solution.

So, we conclude that T_k must be at the bottommost level of the binary tree. The sequence of sub-trees $T_1, T_2, \ldots, T_k$ gives the optimal solution.

In this section we apply backtracking principle to solve the knapsack problem and if necessary, we will look for better principles. In the knapsack problem, we are given n objects, a knapsack of size (capacity) C, a global array $v[1\ldots n]$ giving the values of objects and a global array $w[1\ldots n]$, giving the weight of each object. For example, let $n = 5$ and $C = 10$.

Sr. No:	1	2	3	4	5
v:	8	2	1	5	12
w:	4	2	2	1	6

Initially, no decision is made whether the first object selected to be placed in the knapsack. The set of objects on which a decision is to be made is {1, 2, 3, 4, 5}. This fact is given as follows:

{ } {1, 2, 3, 3, 4, 5}

We branch to the left if no decision is made on the first object available and if a decision is made, we move to the right.

The left branching shows that no decision is made on the first object and right branching shows that a decision is made on the first object.

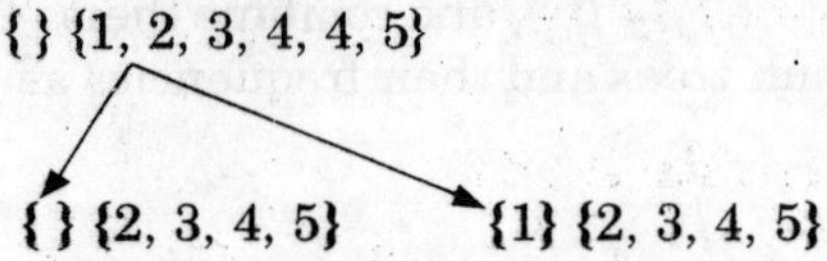

We continue this process. May be at the next step, we get,

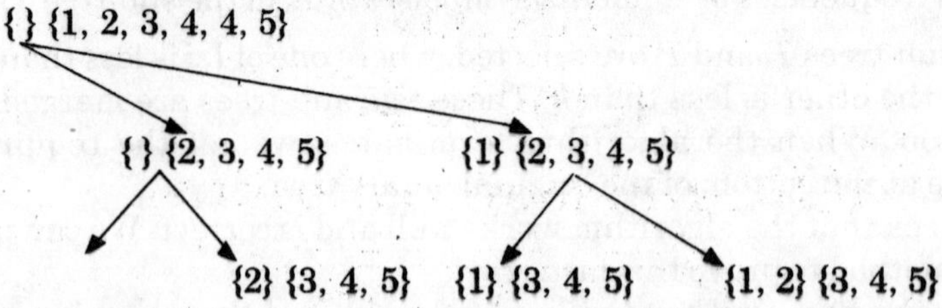

As constructed {2} {3, 4, 5} on the left indicates that the object 1 is not selected and that the object 2 is selected. It results in the value of C as $0 - 2 = 8$.

We now give a recursive procedure for this problem. Let i indicate the index of the first undecided object (this is the object that can put in the knapsack).

```
Ks(C, i)
   {
     if i > n return 0 ←——— I
      if C < w[i] return Ks(C, i + 1) ←——— II
         /* We cannot do anything about the object i. */
         /* Go to object i + 1 */
      else
     return max(Ks(C, i + 1), v[i] + Ks(C − w(i), i + 1) ←——III
   }
```

We apply the aforesaid recursive procedure to the above example of knapsack problem. The questions we ask and the answers we get are shown below:

is i (=1) > n (=5) ? NO

is C (=10) < $w[i]$ (=4) ? NO

So, we apply Ks(10, 2). Continuing this way, we obtain the following at some stage:

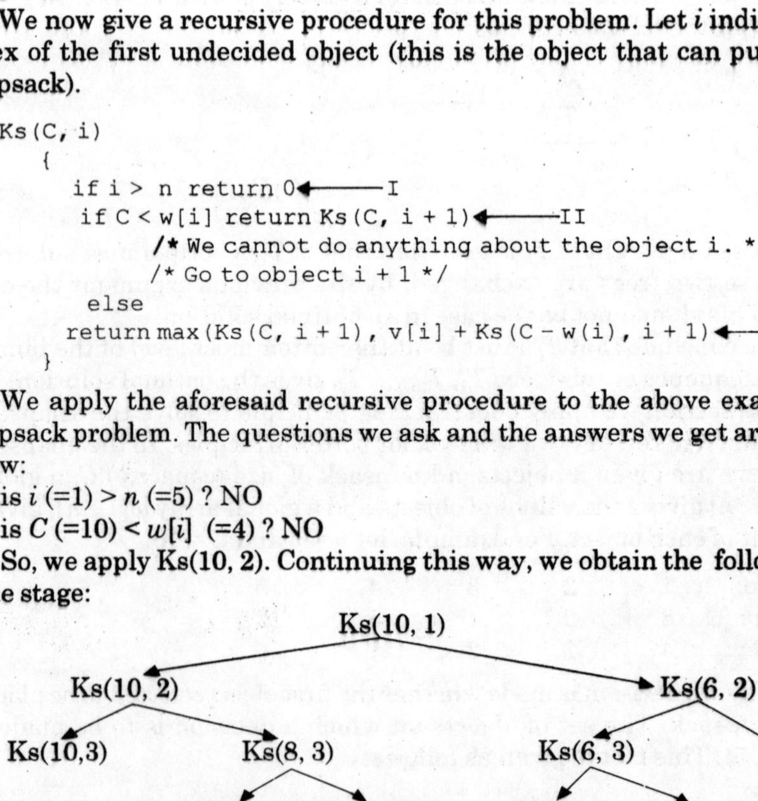

Finally, we obtain a recursive tree. Observe that the recursive call *Ks*(6, 4) is made twice implying that the same set of calculations are redone. The time wasted can be avoided by doing some sort of accounting by maintaining a matrix MAT[c, i], where $1 \le i \le n$ and $0 \le c \le C$. This matrix stores all *Ks*(c, i). A simple check with entries in this matrix will avoid the repetition of calculations. In the recursive procedure for *Ks* (algorithm) given above, we replace the recursive calls I, II, and III as shown below:

```
if i > n
    MAT [c, i] = 0
    return MAT [c, i]
     if MAT [c, i + 1] = null /* No calculations done */
MAT [c, i + 1] = Ks (c, i + 1)
return MAT [c, i + 1]
     if MAT [c, i + 1] = null
MAT [c, i + 1] = Ks (c, i + 1)
return MAT (c, i + 1)
     if MAT (c-w[i], i + 1) = null
return max (MAT [c, i + 1], v[i] + MAT [c - w[i], i + 1])
```

To make things easier, a diary of triplets is maintained for each line. Each triplet consists of line number followed by the available capacity and finally followed by the index of the first available object. It is of the form (ln, c, i). Thus, we can have the following entries in the transcript:

(1, 10, 1)
(2, 10, 1)
....
(1, 10, 2) Observe that this entry corresponds to line 1.

The number of such triplets is exactly equal to the number of time steps in the program. Repeated calculations are avoided by ensuring that the triplets involved are all distinct.

The number of different triplets

= Size of the transcript × number of lines in the program (number of different values i can take) × maximum capacity of the knapsack

= $O(1) * n * C$

= $O(nC)$ and this is far less than 2^n

Thus the time complexity when recursion is used is much smaller than case of backtracking.

Example 3.1

Consider the following instance of the knapsack problem: $n = 3$, $m = 40$, (p_1, p_2, p_3) = (50, 48, 30), and (w_1, w_2, w_3) = (36, 30, 20).

Following are four feasible solutions of this problem:

(x_1, x_2, x_3)	$\Sigma w_i x_i$	$\Sigma p_i x_i$
1. (1/2, 1/3, 1/4)	33	48.5
2. (1, 2/15, 0)	40	56.4
3. (0, 2/3, 1)	40	62
4. (0, 1, 1/2)	40	63

Of these four feasible solutions, we observe that solution 4 gives the maximum profit.

We now give the necessary algorithm for the knapsack problem.

Algorithm 3.2 Knapsack problem–Greedy method

```
Algorithm KnapsackGreedy (m, n)
// The arrays p[1 : n] and w[1 : n] give the profits and weights of n
// objects respectively. m is the capacity of the knapsack.
// We suppose that the n objects are so ordered that for all
possible i,
// p[i] / w[i] ≥ p[i+1] / w[i+1]. x[1 : n] is the solution vector
{
  for j := 1 to n do
    x[j] := 0.0;
  tc := m ;
    for j := 1 to n do
    {
      if (w[j] > tc ) then break ;
      x[j] := 1.0 ;
      tc := tc - w[j] ;
    }
    if ( j ≤ n ) then
      x[j] := tc / w[j] ;
}
```

3.3 TREE VERTEX SPLITTING PROBLEM (TVSP)

$T = (V, E, w)$ is a weighted directed tree. $w(i, j)$ is the weight of the edge $(i, j) \varepsilon E$. When there is no edge between the nodes i and j, we set $w(i, j)$ to a very large number. For any path P from node i to node j, its delay, $d(P)$ is defined as the sum of the weights in that path. If T is a tree then the delay of the tree $d(T)$ is the maximum of all path delays.

Let X be a subset of V. We split each vertex u into nodes u' and u'' obtaining a forest T/X. Each edge (u, j) is split into two edges (u'', j) and

(j, u'). This implies that any outgoing edge from u leaves from u'' and any incoming edge from u enters at u'. Figure 3.1 shows a tree before and after splitting the node 3.

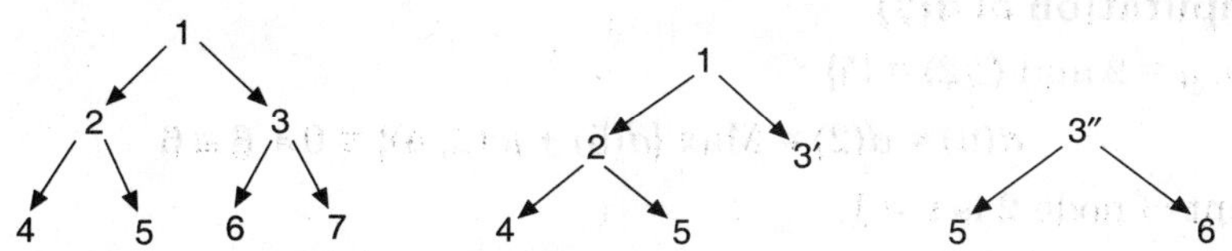

Figure 3.1 A directed binary tree before and after splitting the node 3.

The Tree Vertex Splitting Problem is to get a subset X, of V, containing minimum number of nodes such that $d(T/X) \le \delta$, for a given tolerance limit δ. Given a subset X of V, we can compute $d(T/X)$ in $O(|V|)$ time. One simple and straightforward way of solving TVSP is to enumerate all the subsets of V and compute $d(T/X)$ for each subset X. This method takes a lot of time and as such is not suggested, as there are as many as $2^{|V|}$ subsets (even if the tree contains just 5 vertices, then $2^{|V|} = 2^5 = 32$. large !). A better algorithm is given by the greedy method, by noting that X has minimum number of nodes.

Let $u \varepsilon V$. We compute the maximum delay $d(u)$ from u to any other node in the subtree of u. We split u into two nodes as described above, if it has a parent v such that $d(u) + w(v, u) > \delta$. After this is done $d(u)$ is set to 0. Consider the following directed binary tree:

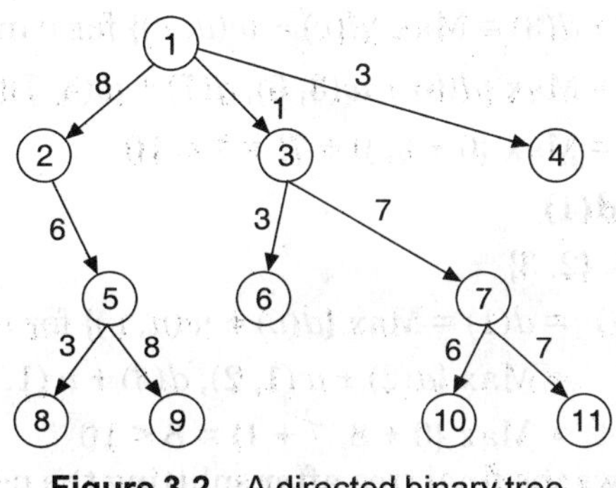

Figure 3.2 A directed binary tree.

In Figure 3.2, we set the tolerance limit $\delta = 10$. The delay for any leaf node is set to zero. Thus,

$$d(4) = d(8) = d(9) = d(6) = d(10) = d(11) = 0$$

We now compute the delays of all non-leaf nodes. If u is an internal node, and $C(u)$ is the set of its children then $d(u)$ is given by

$$d(u) = \max \{d(v) + w(u, v)\}, \text{ maximum over all } v \ \varepsilon\ C(u)$$

Computation of d(5)

Here $u = 5$ and $C(u) = C(5) = \{8, 9\}$. Therefore,

$$d(5) = \text{Max} \{d(8) + w(5, 8), d(9) + w(5, 9)\} = \text{Max} \{0 + 3, 0 + 8\} = 8$$

Parent of the node 5 is $v = 2$.
Since $d(u) + w(v, u) = d(5) + w(2, 5) = 8 + 6 = 14 > 10$, the node 5 is split into two nodes. We set $d(5) = 0$.

Computation of d(2)

Here, $u = 2$ and $C(2) = \{5\}$

$$d(u) = d(2) = \text{Max } \{d(5) + w(2, 5)\} = 0 + 6 = 6$$

Parent of node 2 is $v = 1$.

$$d(u) + w(v, u) = d(2) + w(1, 2) = 6 + 8 = 14 > 10$$

Therefore node 2 is split into two nodes and we set $d(2) = 0$.

Computation of d(7)

Here, $u = 7$ and $C(u) = C(7) = \{9, 10\}$

$$\begin{aligned} d(u) &= d(7) = \text{Max } \{d(v) + w(u, v)\} \text{ for } v \text{ in } \{9, 10\} \\ &= \text{Max } \{d((9) + w(7, 9), d(10) + w(7, 10)\} \\ &= \text{Max } \{0 + 6, 0 + 7\} = 7 \end{aligned}$$

Also, the parent of node 7 is $v = 3$

$$d(u) + w(v, u) = d(7) + w(3, 7) = 7 + 7 = 14 > 10$$

Therefore, node 7 is split into two nodes and we take $d(7) = 0$.

Computation of d(3)

Here, $u = 3$ and $C(u) = C(3) = \{6, 7\}$

$$\begin{aligned} d(u) &= d(3) = \text{Max } \{d(v) + w(u, v)\} \text{ for } v \text{ in } \{6, 7\} \\ &= \text{Max } \{d(6) + w(3, 6), d(7) + w(3, 7)\} \\ &= \text{Max } \{0 + 3, 0 + 7\} = 7 < 10 \end{aligned}$$

Computation of d(1)

$u = 1$, $C(u) = C(1) = \{2, 3\}$

$$\begin{aligned} d(u) &= d(1) = \text{Max } \{d(v) + w(u, v)\} \text{ for } v \text{ in } \{2, 3\} \\ &= \text{Max } \{d(2) + w(1, 2), d(3) + w(1, 3)\} \\ &= \text{Max } \{0 + 8, 7 + 1\} = 8 < 10 \end{aligned}$$

Figure 3.3 shows the final tree after splitting the nodes 2, 5 and 7.

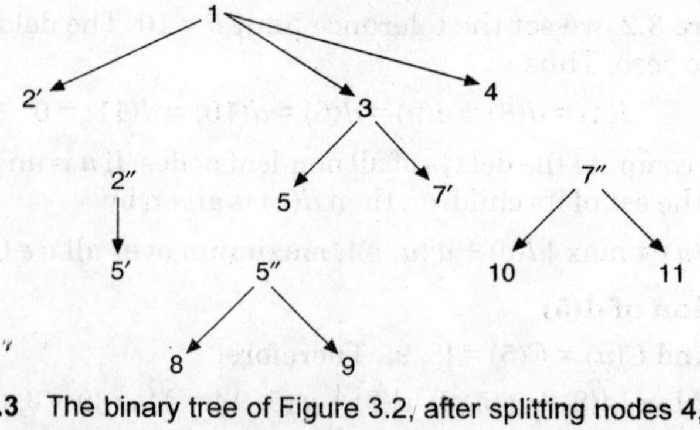

Figure 3.3 The binary tree of Figure 3.2, after splitting nodes 4, 2 and 6.

We now give a tree vertex splitting algorithm. This algorithm is first invoked by calling TVS(Root, δ). Nodes are visited in the "post order tree traversal way".

Algorithm 3.3 The tree vertex splitting algorithm

```
Algorithm TVS (tree, δ)
// This algorithm identifies and gives the nodes to be split in T.
//d[] stores the delay in a node. δ is the tolerance limit.
  {
      if (tree ≠ 0) then                          //tree is not empty
        {
           d(tree) := 0 ;                             // Initialise
           for each child u of tree do
           {
              TVS[u, δ] ;
              d(tree) := max{d(tree), d[u] + wt(tree, u) ;
           }
if ((tree≠root) and (d(tree) + wt(parent(tree), tree) > δ then
              {
                 write (tree);
                 d(tree) := 0 ;
              }
           }
  }
```

3.4 JOB SEQUENCING PROBLEM WITH DEADLINES

There are n jobs. Each job is to be processed for one unit of time, on the only one machine available. For any job i, the non-negative deadline $d(i)$ is an integer greater than or equal to zero. A profit $p(i)$ is earned for job i if and only if it is completed within its deadline. Now let the array $JS[\,]$ consist the indices of jobs that can be completed within their deadlines. JS is then a feasible solution for this problem. The value of a feasible solution JS is the sum of the profits of jobs in JS, or Σp_i for $i \in JS$. A feasible solution with maximum value is an optimal solution. This problem is again an example of subset paradigm as it involves identification of subsets of a set.

Example 3.2

Let $n = 4$, $(p_1, p_2, p_3, p_4) = (200, 20, 30, 55)$ and $(d_1, d_2, d_3, d_4) = (2, 1, 2, 1)$. The feasible solutions and their values are tabulated below:

	Feasible solution	*Processing sequence*	*Value*
1	(1, 2)	2, 1	220
2	(1, 3)	1, 3 or 3, 1	230
3	(1, 4)	4, 1	255
4	(2, 3)	2, 3	50
5	(3, 4)	4, 3	85
6	(1)	1	200
7	(2)	2	20
8	(3)	3	30
9	(4)	4	55

From the above we observe that solution 3 is optimal. In this optimal solution jobs 1 and 4 only are processed in the order 4, 1 and the maximum profit is 255. The processing of job 4 begins at time zero and is completed in one hour and then the processing of job 1 begins. This job is completed at the end of time 2 units. The following theorem gives a necessary and sufficient condition for a feasible solution of this problem.

Theorem: Let *JS* be a set of *k* jobs and $\sigma = i_1, i_2, \ldots, i_k$ a permutation of jobs in *JS* such that $d_{i1} \le d_{i2} \le \ldots \le d_{ik}$. Then *JS* is a feasible solution if and only if the jobs in *JS* can be processed in the order σ without violating the deadline.

We now give the job sequencing with deadlines in algorithm 3.4. The original sequence of jobs is re-altered so that $p[1] \le p[2] \le \ldots \le p[n]$. Array *JS* denotes the optimal solution. *JS* is initialised by including the first job from the original ordered list of jobs. Then the *k*th job ($2 \le k \le n$) is considered and is tested for its feasibility to be included in *JS*.

Algorithm 3.4 Job sequencing

```
Algorithm Job-Seq(d, 1, n)
// Algorithm for Job Sequencing with dead lines d[].
{
    d[0]: = 0;
    JS[1]: = 1;
    i: = 1;
    for j: = 2 to n do
      {
        k: = i;
        while ((d[JS[k] > d[j]) and (d[JS[k]] ≠ k)) do
        k: = k-1;
          if ((d[JS[k]]≤d[j] and (d[j] > k)) then //Testing for
          feasibility
          {
          for m: = i to (k+1) step -1 do
```

```
                JS[m+1]:=JS[k];
                JS[k+1]:=j;
                i:=i+1;
            }
        }
    return i;
}
```

It is established that the computing time of job sequencing can be reduced from $O(n^2)$ to $O(n)$.

3.5 MINIMUM-COST SPANNING TREES

Definition: Let $G = (V, E)$ be an undirected graph. A subgraph $T = (V, E')$ of G is a spanning tree of G if and only if T is a tree.

Example 3.3

Figures in 3.4(a) and (b) below represent a graph and its spanning tree.

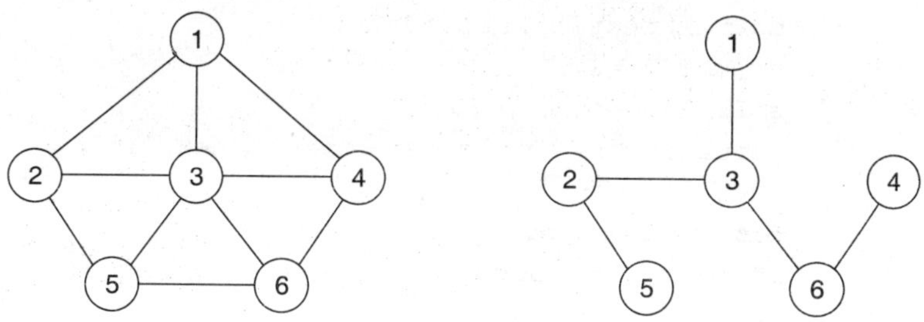

(a) (b)

Figure 3.4 A graph and spanning tree of an undirected graph.

3.5.1 Prim's Algorithm

$G = (V, E)$ is a given weighted graph. Let $V = \{1, 2, \ldots, n\}$, the *minimum spanning tree of G is given in Algorithm (3.5)*. That is, it computes the subset U of V that contain the edges of the minimum spanning tree. To begin with we take $U = \{1\}$. Then U grows one edge at a time. An edge (i, j) is considered and a selection criterion is applied to decide whether to include it in U or not. We ensure that U remains a tree at every stage. The best selection criterion is to select an edge (i, j) that contribute minimum cost to the already accumulated partial minimum cost. To locate such an edge (i, j) with the vertex i in the tree and the vertex j not yet in U, we consider the set $S(j)$ = {all children of node j}. Among them we take the vertex near(j) so that the cost of the edge (j, near(j)) is minimum. We set 0 to near(j) if j is included in the tree.

For this, the greedy method is applied. The Prim's algorithm 3.5 is given below.

Algorithm 3.5 Prim's algorithm

```
1.  Algorithm Prim(E, cost, n, t)
2.  // Identifies the minimum spanning tree.
3.  // The minimum spanning tree if exists is stored
4.  // in the array st[1 : n-1, 1 : 2]
5.    {
6.      Let (i, j) be an edge of minimum cost in E ;
7.      totmincost: = cost[i, j];
8.      st[1, 1]: = i;
9.      st[1, 2]: = j;
10.     for k: = 1 to n do // Initialisation of next
11.     if cost[k, l] < cost[k, i] then next[k]: = j;
12.     else next[k]: = i;
13.     next[i]: = 0;
14.     next[j]: =0;
15.     for k: = 2 to n-1 do
16.       Let l be an index such that next[l] ≠ 0 and
17.       cost[l, next[l]] is minimum;
18.       st[k, 1]: = l;
19.       st[k, 2]: = next[l];
20.       totmincost: = totmincost + cost[l, next[l]];
21.       next[l]: = 0;
22.       for i: = 1 to n do // Update next[ ].
23.       if (next[i] ≠ 0) and (cost[i, next[i]] > cost[i, l])) then
          next[i]: = l;
24.   } return totmincost; }
```

It is to be noted that a spanning tree is a minimal subgraph (a graph having minimum number of edges) G' of G such that $V(G) = V(G')$ and G' is connected.

Time Complexity of Prim's Algorithm

In view of the two for loops in steps 20 and 28 in the above algorithm, we observe that the total time of Prim's algorithm is $O(n^2)$. (Some other steps take $O(n)$ or $\Theta(1)$ time and as such, we need not bother about these times).

3.5.2 Kruskal's Algorithm

There is another algorithm for finding the minimum spanning tree of a given weighted graph G and this one is called the Kruskal's algorithm. For applying this algorithm, a list of all edges of G in the non-decreasing order is maintained. But this is similar to the Prim's algorithm. As in Prim's algorithm, the set U of edges that are selected so far leads to tree. Following is a general form of Kruskal's algorithm.

1. temp = φ
2. repeat up to step 6 till (Number of elements (edges) in temp < $n-1$) and ($E \neq \varphi$)
3. Select an edge (i, j) minimum cost
4. Delete (i, j) from E
5. If inclusion of the edge (i, j) into temp does not create a cycle in temp then add (i, j) to temp
6. else discard (i, j)
7. Stop

We now give the Kruskal's algorithm.

Algorithm 3.6 Kruskal's algorithm

```
Algorithm Kruskal (E, cost, n, t)
// Kruskal's Algorithm for identifying the minimum spanning
tree
{
   Construct a heap out of the edge costs.
for k: = 1 to n do
   parent[k]: = -1;
   k: = 0;
   totmincost: = 0.0;
      while ((k < n-1) and (heap not empty)) do
         {
            Delete a minimum cost edge (u, v) from the heap and
            reheap the remaining edges;
            l: = Find(u); m: = Find(v);
            if (l ≠ m) then
            {
               k: = k+1;
               temp[k, 1]: = u;
               temp[k, 2]: = v;
               totmincost: = mincost + cost[u, v];
               Union(l, m);
            }
         }
         if (k ≠ n-1) then write (" No spanning tree");
         else return mincost;
}
```

It is seen that Kruskal's algorithm takes $O(|E| \log |E|)$ time.

3.6 OPTIMAL STORAGE ON TAPES

Suppose n computer programs are to be stored on a magnetic tape of length l. Suppose that the length of program i is l_i, for $1 \le i \le n$. If the sum of the lengths of all programs is at most l, then all the programs can be stored on the tape without any omissions. We suppose that the tape is positioned at the beginning of the tape. As such if the programs are stored in the order $I = i_1$, $i_2, \ldots, i_n$, then the time t_j, needed to retrieve program ij is proportional to

$$\Sigma l_{i_k} = l_{i_1} + l_{i_2} + \ldots + l_{i_k}, \text{ for } 1 \le k \le j$$

The expected or mean retrieval time (MRT) is $(1/n)\ \Sigma t_j$, for $1 \le j \le n$.

"The optimal storage problem is to find a permutation for the n programs so that when they are stored on the tape in this order the MRT is minimized". This problem amounts to the ordering paradigm. Minimizing the MRT amounts to minimizing

$$d(I) = \Sigma\Sigma l_{i_k}, \quad \text{for} \quad 1 \le j \le n, 1 \le k \le j$$

We first illustrate this process in the following example.

Example 3.4

Let $n = 3$. $(l_1, l_2, l_3) = (8, 12, 2)$. As $n = 3$, there are $3! = 6$, possible orderings. All these orderings and their respective d values are given below:

Ordering	$d(i)$	
1, 2, 3	8 + (8 + 12) + (8 + 12 + 2)	= 50
1, 3, 2	8 + 8 + 2 + 8 + 2 + 12	= 40
2, 1, 3	12 + 12 + 8 + 12 + 8 + 2	= 54
2, 3, 1	12 + 12 + 2 + 12 + 2 + 8	= 48
3, 1, 2	2 + 2 + 8 + 2 + 8 + 12	= 34
3, 2, 1	2 + 2 + 12 + 2 + 12 + 8	= 38

The optimal ordering is 3, 1, 2.

The greedy method is now applied for solving this problem. It requires that the programs are stored in non-decreasing order which can be done in $O(n \log n)$ time. This method selects some suitable optimization criterion. Any increase in the existing length of the tape is to be minimized. That is, if we have already constructed the permutation $i_1, i_2, \ldots, i_r$, the next coming program j gives the permutation $i_1, i_2, \ldots, i_r, i_{r+1} = j$. The value of d is increased by $\Sigma\ li_k + l_j$, for $1 \le k \le r$.

It is to be observed that the MRT is minimized when programs are stored in non-decreasing order of their lengths.

Theorem: If $l_1 \le l_2 \le \ldots \le l_n$, the ordering $i_j = j$, $1 \le j \le n$, minimizes

$$\Sigma\Sigma\ l_{ij} \text{ for } 1 \le k \le n,\ 1 \le j \le k.$$

overall possible permutations of the i_j.

Algorithm 3.7 Assigning programs to tapes

```
1. Algorithm Store(n, m)
2. // Optimal storage of n programs on m tapes.
```

```
{
   k: = 0: // Next tape to be stored
   for j: = 1 to n do
   {
     write ("assign program", j, "to tape", k);
     k: = (k + 1) mod m;
   }
}
```

3.7 SINGLE SOURCE SHORTEST PATHS

$G = (V, E)$ is a weighted, directed graph. The problem is: "to determine the shortest paths from a given source vertex v_1 to all the remaining vertices of G". We observe that this problem is an ordering paradigm, as we order subsets to find shortest path from a source vertex v_1 to a destination vertex v.

Example 3.5

Consider the weighted, directed graph as shown in Figure 3.5.

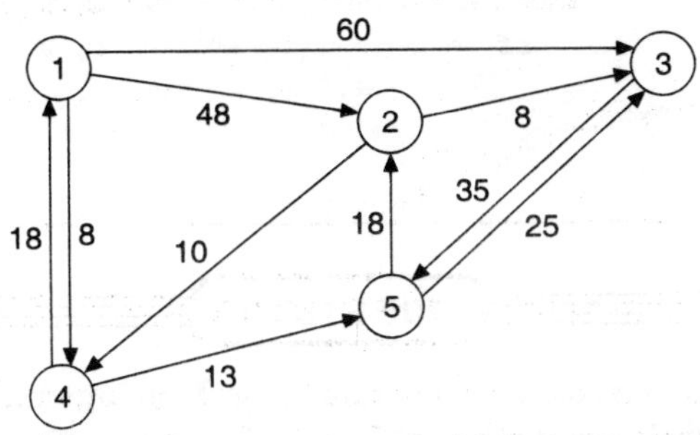

Figure 3.5 A weighted Graph.

The shortest paths from vertex 1 are listed below.

From Node	*To Node*	*Shortest Path*	*Shortest Length*
1	2	1, 4, 5, 2	39
1	3	1, 4, 5, 3	46
1	4	1, 4	8
1	5	1, 4, 5	21
1	6	No path	

Greedy method is applied to solve this problem. A suitable optimization measure is to be applied, so as to make the sum of the lengths of all paths so far generated a minimum.

We now give the necessary algorithm.

Algorithm 3.8 A greedy algorithm for single source shortest paths

```
Algorithm Single-Source-ShortestPaths(u, cost, d, n)
//G(V, E) is given by its cost adjacency matrix cost[1:n, 1:n]
//v is the given source vertex. Length of a path d[i] from u to
//a vertex j is defined as the sum of the costs of edges on the path
//to vertex j in G. SP[] is a Boolean function and SP[j] returns
//"true" if node j is in the shortest path, otherwise it returns
"false"
{
    for j:= 1 to n do
      {
        SP[u]: = true; // Include u in SP
        d[j]: = cost[u, j];
      }
        SP[u]: = true;
        d[v]: = 0.0;
        for i: = 2 to n-1 do
        {
          Choose vertex v such that d[u] is minimum, not in SP
            SP[v]: = true; // Include v in SP.
            for (each w adjacent to v with SP[w] = false) do
              if (d[w] > d[v] + cost(v, w])) then
                d[w]: = d[u] + cost[u, w];
        }
}
```

QUESTIONS

1. What is the knapsack problem? Find an optimal solution to the knapsack problem with $n = 7$, $m = 15$, $(p_1, ..., p_7) = (10, 5, 15, 7, 6, 18, 3)$ and $(w_1, ..., w_7) = (2, 3, 5, 1, 4, 1)$.
2. Write an algorithm for the optimal solution of the knapsack problem.
3. Explain in detail the tree vertex splitting problem.
4. Write an algorithm for the tree vertex splitting problem.
5. Explain the job sequencing problem.
6. Write an algorithm for the job sequencing problem.
7. Solve the following job sequencing problems:
   (1) $n = 5$, $(p_1, ..., p_5) = (20, 15, 10, 5, 1)$ and $(d_1, ..., d_5) = (2, 2, 1, 3, 3)$.
   (2) $n = 7$, $(p_1, ..., p_7) = (3, 5, 20, 18, 1, 6, 30)$ and $(d_1, ..., d_7) = (1, 3, 4, 3, 2, 1, 2)$.
8. Solve the job sequencing problem with $p[1, 4] = [100, 10, 15, 27]$ and $d[1.4] = [2, 1, 2, 1]$.

9. Define a minimum-spanning tree. Write Prim's algorithm and discuss its time complexity.

10. Write Kruskal's algorithm and discuss its time complexity.

11. Explain the "Optimal storage on tapes" problem.

12. Write a greedy algorithm for the shortest path problem.

13. Construct a minimal spanning tree for the tree with the following graph:

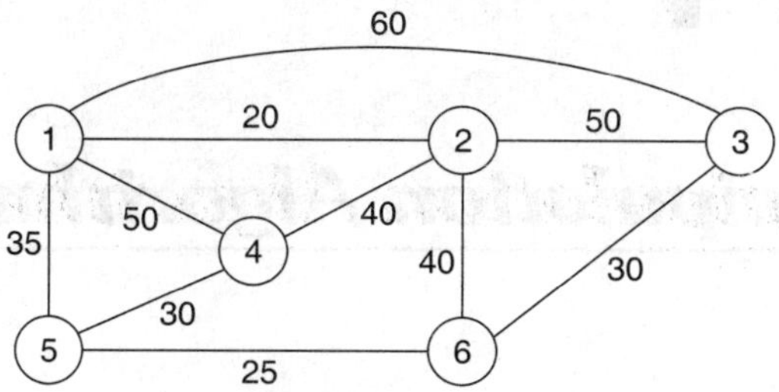

14. Explain the "Greedy Method". Write an algorithm for the control abstraction of this method.

CHAPTER 4

Set Manipulation Algorithms

4.1 INTRODUCTION

In this chapter we consider the different operations on sets, binary search trees, 2–3 trees, AVL trees and some aspects of heaps.

4.2 SET OPERATIONS

For convenience, we give indices to elements of a set and consider its elements as 1, 2, ... Let us assume that the sets involved are pairwise disjoint. Consider the set $S = \{1, 5, 6, 10\}$. One way of representing a set is to make one of its elements a parent and the remaining elements its children and link the children to the parent (generally the parent is linked to the children). This representation is shown below in Figure 4.1.

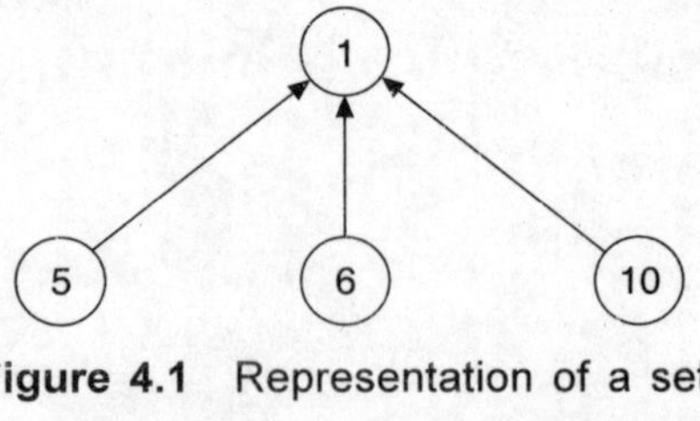

Figure 4.1 Representation of a set.

1. MEMBER (i, S): This is a boolean function and returns 'yes' (or 1) if i is a member of S otherwise, it returns 'no' (or 0). For example, if $i = 5$ and $S_1 = \{1, 5, 6, 10\}$, then MEMBER (5, S_1) returns 1 and if $i = 7$, then MEMBER (i, S_1) returns 0.

2. INSERT (i, S): Adjoins the element i into S. In the above example, INSERT (8, S_1) returns the set S = {1, 5, 6, 10, 8}. That is INSERT (i, S) returns S U {i}.
3. DELETE (i, S) returns S – {i}.
4. UNION (S_1, S_2, S_3) returns $S_3 = S_1$ U S_2. To get the representation of S_1 U S_2, we set the parent field of one of the roots to the other root. Suppose S_1 = {1, 5, 6, 10} and S_2 = {2, 4, 9}. Then union of sets S_1 and S_2, is either of the following:

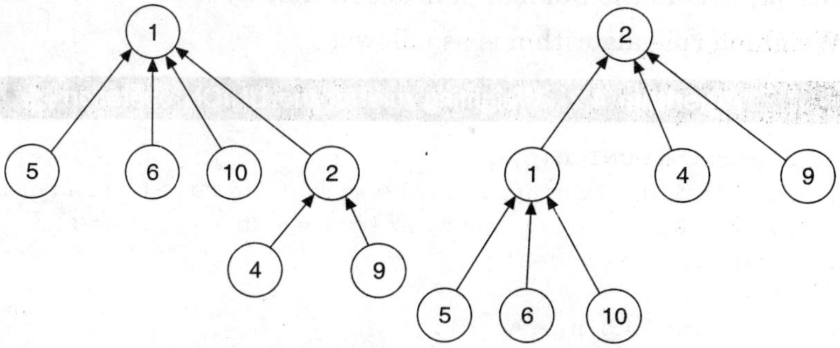

Figure 4.2 Two possible ways of S_1 U S_2.

5. Find (i): Finds the set containing i.
   Find (5): Gives the set S_1 = {1, 5, 6, 10}

We introduce an array $p[1 \ldots n]$ so that each element of this array p is a representative of a tree. For example, the kth element of this array represents the tree node that contains the element k. If k happens to be a root node, then $p[k]$ returns –1. With the sets S_1, S_2 and S_3 taken as above, we have

$$p[4] = 2, p[2] = -1, p[6] = 1, \text{ etc.}$$

Necessary algorithms for Union and Find functions can be written as shown below in terms of the array p:

Algorithm 4.1 Algorithms for "Union" and "Find" functions

```
Algorithm SUnion(i, k)
// This is an algorithm to implement the "union" and "find"
functions.
{
  p[i]: = k;
}
```

```
Algorithm SFind(k)
{
  while (p[k] ≥ 0) do
    k: = p[k];
  return k;
}
```

Exercise

Show that the time complexity of the Find operation is $O(n^2)$.

Note that the time complexity of the Find operation is too large. This can be reduced using the following rule for UNION (i, j).

Definition [Weighting rule for Union (i, j)]: "If the number of nodes in the tree with root i is less than the number in the tree with root j, then make j the parent of i; otherwise make i the parent of j". If k is a node, let Nnodes $[k]$ denote the number of nodes in that tree.

Weighing rule algorithm is as follows:

Algorithm 4.2 Weighting rule for the UNION operation

```
AlgorithmWUNION(p, q)
// This algorithm applies the weighting rule to find union of
// sets with roots p and q. Ap[p]: = - Nnodes[p] and
// Ap[q]: = - Nnodes[q].
  {
     sum: = Ap[p] + Ap[q];
     if (Ap[p] > Ap[q]) then
        { //p has less number of nodes.
           Ap[p]: = q;
           Ap[q]: = sum;
        }
     else
        { // q has less number of nodes.
           Ap[q]: = p;
           Ap[p]: = sum;
        }
  }
```

It is observed that the above reduced time complexity can be further reduced by making small changes in the algorithm by applying what is called the "**collapsing rule**" as defined below.

Definition [Collapsing Rule]: "If node j is on the path from i to its root and $p[i] \neq$ Root$[i]$, then set $p[j]$ to Root$[i]$".

Following is the algorithm for the implementation of the collapsing rule.

Algorithm 4.3 Collapsing rule

```
AlgorithmCFind(k)
// This algorithm finds the root of the tree containing
// the element k, using collapsing rule. All nodes
// from k to the root are collapsed.
  {
```

```
        root: = k;
        while (Ap[root] > 0) do
           root: = Ap[root];
        while (k ≠ root) do
        {
           j: = p[k];
           p[k]: = root;
           k: = j;
        }
        return root;
    }
```

4.3 TREES

A tree is recursively defined as a non-empty, finite set of elements called nodes, which contains a special node called the root node and the remaining nodes are partitioned into some $m \geq 0$ disjoint sets $T_1, \ldots, T_m$, where each T_i is a tree.

By the **degree of a node**, we mean the number of its subtrees. In the tree of Figure 4.3, degree of node *A* is 3, degree of node *F* is 2 and that of node *E* is zero. **Leaf nodes**, otherwise called terminal nodes are those whose degree is zero. In the above tree, leaf nodes are *E, L, M, G, N, I, J* and *P*. All other nodes are called **internal nodes or non-terminal nodes**. The roots of the subtree of a node *N* are called the **children** of *N*. Node *N* is called the parent of the children. Children of the same parent are called siblings. By the **level** of a node, we mean the following: Level of the root node is 1. The level of children of a node with level *k* is *k* + 1. The **height (or depth)** of a tree is the maximum level of any node in the tree. A **forest** is a set of nodes greater than or equal to zero disjoint trees.

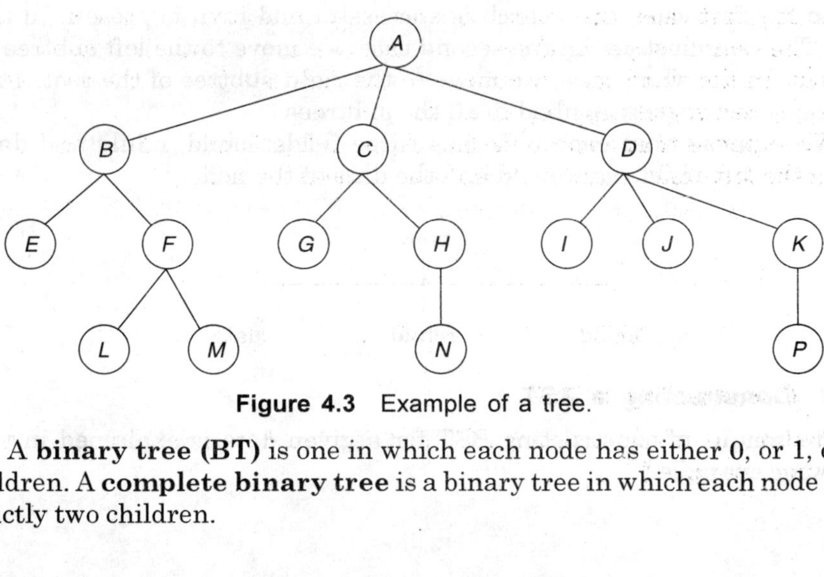

Figure 4.3 Example of a tree.

A **binary tree (BT)** is one in which each node has either 0, or 1, or 2 children. A **complete binary tree** is a binary tree in which each node has exactly two children.

4.4 DICTIONARIES

When a set is used in designing an algorithm, what we want to know is the present elements of that set. A dictionary is a data structure in which operations like Member, Insertion, Deletion and Search operations can be performed (in general these operations are to be performed frequently).

We now consider special trees called binary search trees. Every node is given a label (key). Suppose that no two labels are equal.

4.5 BINARY SEARCH TREE (BST)

A Binary Search Tree is a special binary tree (may be empty) that satisfy the following properties:

1. The labels in the left subtree of a node are less than the label of its parent.
2. The labels in the right subtree of a node are greater than the label of its parent.
3. The left and right subtrees are also binary search trees.

4.5.1 Searching a BST

A BST and a search key "item" are given. We search whether the search key item is present in the given BST. We begin our search from the root node of the BST. If BST is null, then the search stops. Otherwise, we compare item with the key of the root node of the BST. Following three cases arise:

1. item = key of the root;
2. item < key of the root; or
3. item > key of the root.

In the first case, the search is successful and item is present in the BST. The search stops. In the second case, we move to the left subtree of the root. In the third case, we move to the right subtree of the root. This process is recursively applied to all the subtrees.

We suppose that each node has three fields: lchild, rchild and data giving the left child, right child and the data of the node.

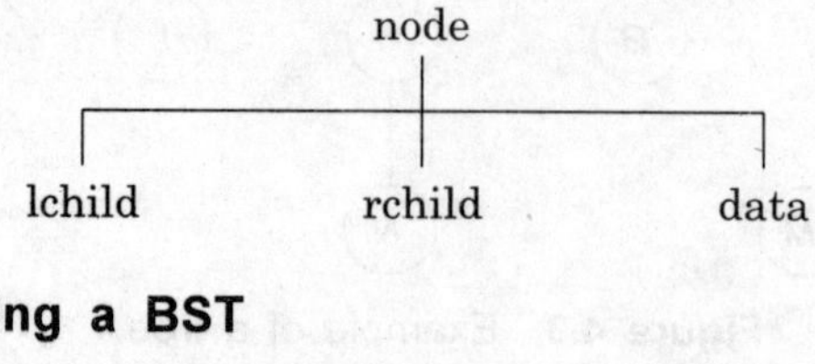

4.5.2 Constructing a BST

The technique of constructing BST for a given data is explained in the following example.

Example 4.1

Construct a BST with the data: 75, 115, 95, 61, 105 and 17.

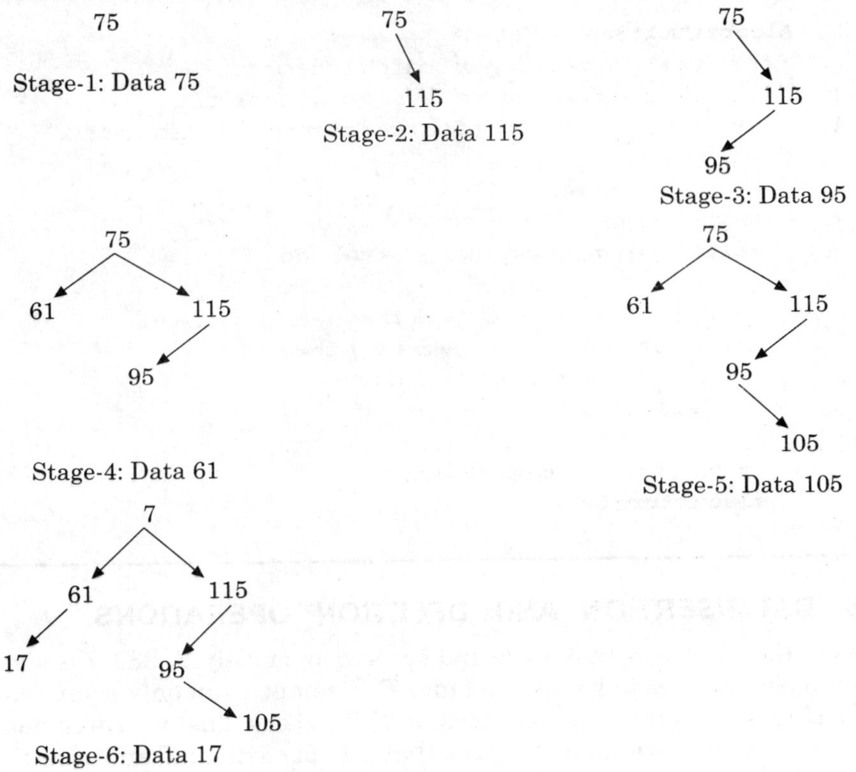

Figure 4.4 Constructing a BST.

Following is the recursive search algorithm of a binary search tree.

Algorithm 4.4 Searching a BST—Recursive approach

```
Algorithm RSearch (bst, item)
// bst is the given binary search tree and
// item is the given search key.
// This algorithm searches the bst for the presence of the key.
  {
     if (bst = 0) then return 0; // The bst is null
   else if (item = bst → data) then return bst;
       else if (item < bst → data) then
           return RSearch (bst → lchild, item);
       else return RSearch (bst → rchild, item);
  }
```

The following is the iterative algorithm for the above search.

Algorithm 4.5 Searching a binary search tree—Iterative approach

```
Algorithm ItSearch (item)
// bst is the given binary search tree and
// item is the given search key. This algorithm
// searches iteratively for the presence of item in bst.
{
   search: = false;
   bst: = tree;
   while ((bst ≠ 0) and (not search)) do
   {
      if (item = (bst → data)) then search := true;
      else if (item < (bst → data)) then
           bst: = (bst → lchild);
        else bst: = (bst → rchild);
   }
   if (not search) then return 0;
   else return bst;
}
```

4.6 BST-INSERTION AND DELETION OPERATIONS

We use the words element, node and key synonymously. A BST T is given and an element x is to be inserted into T. Element x can only be inserted only if it is different from the keys of T. To check this, we carry out a search. Search is successful means that x is present in T and insertion cannot be done. On the other hand, if the search is unsuccessful, the search terminates at a node. Then x is inserted at the point where the search terminated. This process is shown in the Figure 4.5.

Figure 4.5(a) below is a given BST. Suppose we want to insert a node with key 60. A search is first conducted for the presence of the node with key 60. The search is unsuccessful. The last node visited has key 50. So, the new key 60 is inserted as a right child of this node. As a result of this insertion, we get Figure 4.5(b). When we insert node with key 40, we get Figure 4.5(c). In this connection, we note that if a node has a leftsize field, then it is updated. Inserting a node into a given BST is shown below.

4.6.1 Insertion

Algorithm 4.6 Insertion into a binary search tree

```
Algorithm BSTInsert (item)
// This algorithm inserts the node item into the BST, if not
present.
// It is assumed that every node has three fields:
```

```
// lchild, rchild and data.
{
   search: = false;
     t: = tree;
     while ((t ≠ 0 ) and not found) do
       {
          save: = t;
          if (item = (t → data)) then search := true ;
        else if (item < (t < (t → data)) then t:= (t → lchild);
          else t: = (t → rchild);
       }
// Insertion is now made
if (not search) then
   {
     t: = newTNode;
     (t → lchild): = 0;
     (t → rchild): = 0;
     (t → data): = item;
     if (tree ≠ 0) then
       {
          if (item < save → data)) then (save → lchild): = t;
            else (save → rchild): = t;
       }
        else tree: = t;
   }
}
```

4.6.2 Deletion

Consider the following cases:

Case-I: The node to be deleted is a leaf node. Then we set the left or right child field of its parent is set to null (zero), depending upon whether the leaf node is a left or right child of its parent. For example, to delete the node with key 10 in the BST of Figure 4.5(a), we set lchild(par(10)) = 0. We make appropriate change if it were a right child.

Case-II: The node to be deleted has a single child, either a left or a right. Then the node is deleted and in that place its child comes. We use a pointer from the parent node to the single child of the node to be deleted. This process is illustrated below.

Case-III: The node to be deleted is a non-leaf node and two children. This is shown below. If the node to be deleted is the root node of the given BST, then we simply replace the root node either by the largest element in its left subtree or by the smallest element of its right subtree.

The following graphs illustrate the process of inserting nodes with keys 60 (Figure 4.5(b)) and then 40 (Figure 4.5(c)) into the graph (a) of Figure 4.5.

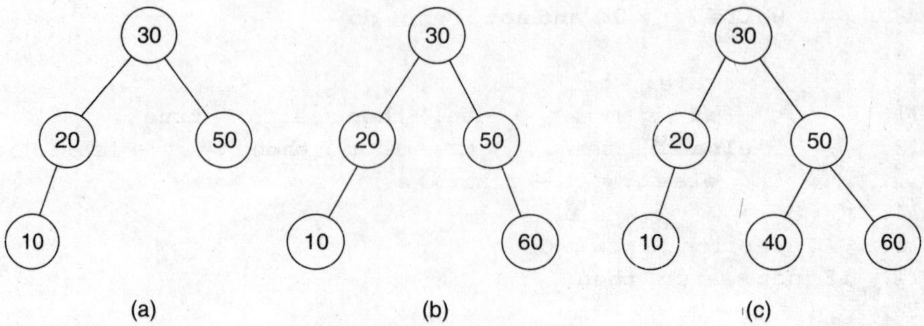

Figure 4.5 Insertion.

In the following, first the root node 35 is deleted from the BST of Figure 4.6(a). As explained in case-III above, it is replaced either by the largest element in its left subtree (15), or by the smallest element in its right subtree (45). Suppose we replace it by 15. Then we obtain the BST of Figure 4.6(b). Now there is a duplication of 15. We delete the second 15, as it has only one child. Finally, we obtain the BST of Figure 4.6(c).

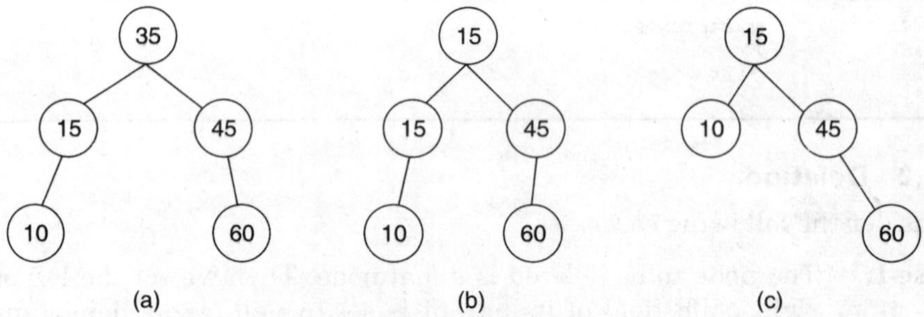

Figure 4.6 Deletion from a BST.

The following "Procedure" identifies, if any, the location, "loc", of a node with a given data "info" in a given BST.

Algorithm 4.7 Finding the location of the node to be deleted in a BST

```
Algorithm Find (data, lchild, rchild, root, info, loc, par)
// info is the given data item of node, if any, to be deleted.
// loc gives location of this node.
// par denotes the parent of this node.
// If loc = NULL and par = Null then the tree is empty and
// deletion is not possible.
// If loc = NULL and par ≠ NULL, then no such node exists in
```

```
// the BST. Deletion is not possible.
// If loc → NULL and par = NULL, then the node to be deleted
// is the root node of the BST.
// Finally, if loc ≠ NULL and par ≠ NULL, then the node with
// required data is present in the BST and deletion is possible.
{
   if root = NULL then // Tree is empty
   {
      loc: = 0;
        par: = 0;
   }
   return;
else if info = data[root] then // Node is the root node
     {
        loc: = root;
        par: = 0;
     }
   return;
else if info < data [root] then // initialize ptr and save
     {
        ptr: = lchild[root];
        save: = root;
     }
else
     {
        ptr: = rchild[root];
        save: = root;
     }
while (ptr ≠ 0) do
   {
       if info = data[ptr] then // search successful
         {
           loc: = ptr;
           par: = save ;
           return ;
         }
       else if info < data[ptr] then
         {
           save: = ptr;
           ptr: = lchild[ptr];
```

```
            }
        else
            {
                save: = ptr;
                ptr: = rchild[ptr];
            }
    } // end of while loop
    if loc = NULL and par ≠ 0 then // search is unsuccessful
        {
            loc: = NULL;
            par: = save;
        }
}
```

The following algorithm deletes a specified node from a BST. The function Find is first invoked to find the location loc of the node to be deleted and its parent par.

Algorithm 4.8 Deletion of a node not having two children

```
Algorithm DeleteBST(data, lchild, rchild, root, loc, par)
// Node n does not have two children: lchild and rchild. LOC is
// the location of the node n to be deleted.
// par gives parent of n.
// par = NULL indicates that the node N is the root node.
// The pointer child gives the location of the only child of n.
// child = NULL indicates that N has no children.
  {
     call Find(data, lchild, rchild, root, loc, par);
// Case of n having no two children.
    if ((lchild[loc] = NULL) and (rchild[loc] = NULL)) then
       child:=NULL;
    else if lchild[loc] ≠ NULL then set child: = lchild[loc];
       else set child: = rchild[loc];
    if par ≠ NULL then
       if loc = lchild[par] then lchild[par]: = child;
          else rchild[par]: = child;
    else root: = child;
    return;
```

4.7 PRIORITY QUEUES

"A priority queue is a data structure that supports the operations of search min (or max), insert min (or max), and delete min (or max)". A priority queue is best represented by an unordered linear list. Suppose that this priority queue supports delete max operation (to get service). For deleting an element from this list one has to first search in the list and then delete if present. As such, each deletion takes Θ(n) time, when there are n elements in the list (unordered). Any insertion is made at the end of time Θ(1).

A better way is to use an ordered linear list. Then the delete time is Θ(1) for each and of O(n) for each insertion. In this connection, we use a data structure, called max heap in which both delete and insert operations can be done in $O(\log n)$ time.

4.8 HEAP AND HEAP SORT

Definition: Let H be a complete binary tree. H is called a MAXHEAP, if each node N of H satisfies the property that the label at N is greater than or equal to those of children of N. A MINHEAP is defined similarly. Figure 4.7 shows a heap:

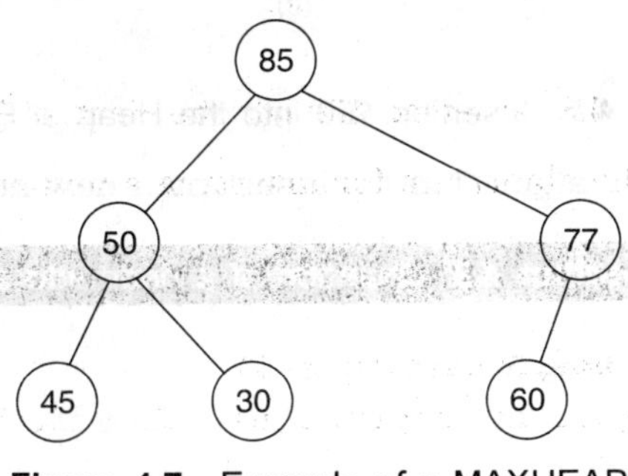

Figure 4.7 Example of a MAXHEAP.

4.8.1 Heap Insertion

A heap H and a value x is given. Now insert a node with label x into H, if it is not already present in H.

1. We first append a node with label x at the end of H. Then H need not be a heap.
2. Next this node with label x is raised to its 'appropriate place' in H, so that H is finally a heap. That is, we reheap H.

This process is shown below. For simplicity, hereafter, we call the node with label x as node or by x itself.

Consider the heap of Figure 4.7. We now insert 100 into it. We first append it at the bottom of the given heap and then reheap it. This process is shown in the following figures.

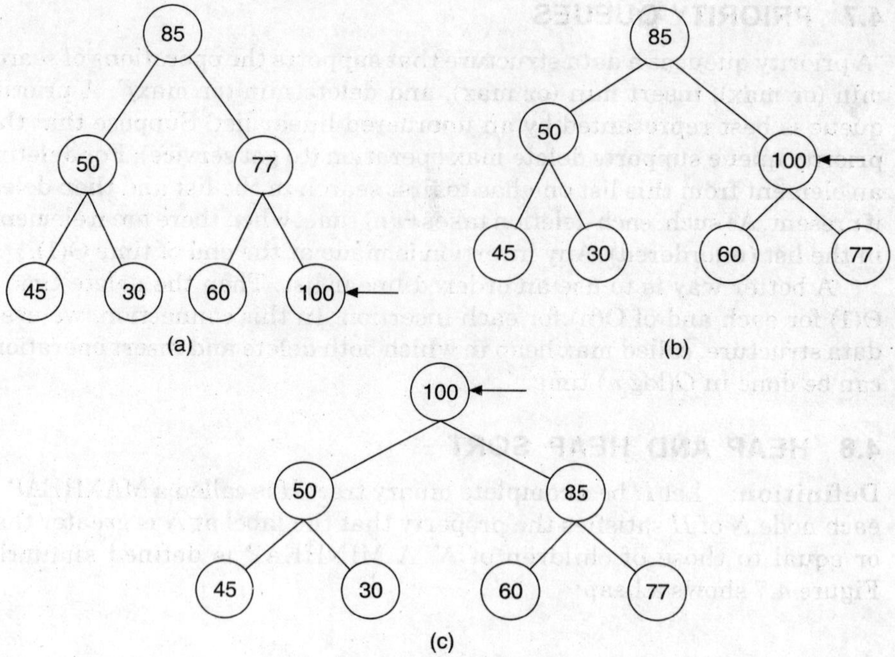

Figure 4.8 Inserting 100 into the Heap of Figure 4.7.

Following is the algorithm for inserting a new node into a heap.

Algorithm 4.9 Insertion of a node into a heap

```
Algorithm HeapInsert (a, n+1)
// This algorithm inserts a[n+1] into the heap with elements
   a[1...n].
{
   k: = n+1;
   x: = a[n+1];
   while ((k > 1) and (a[⌊k/2⌋] < x)) do
      {
         a[k]: = a[⌊k/2⌋];
         k: = ⌊k/2⌋
      }
   a[k]: = x;
   return true;
}
```

It is observed that in the worst case, insertion into a heap with n elements will take Θ(nlogn) time.

4.8.2 Heap Deletion

Deletion of the maximum key from a max heap *H* is now considered. The algorithm DeleteMax (*a*, *n*, *x*) uses algorithm Adjust (*a*, *k*, *n*). In the second algorithm, *a*[1...*n*] is a complete binary tree. If the subtrees at $2k$ and $2k + 1$ are already max heaps, then ReHeap rearranges the elements of a[ ] such that the tree rooted at *k* is also a max heap. The maximum element of the array *a*[ ] happens to be the root of the corresponding complete binary tree. At the end, we call ReHeap (*a*, 1, *n* – 1).

Algorithm 4.10 Deletion from a heap

```
Algorithm DeleteMax (a, n, x)
// This algorithm deletes the maximum element of the heap
// formed by the elements of a[1...n].
{
   if (n = 0) then
      {
         write ("heap is empty");
         return false ;
      }
      x: = a[1]; // saves the maximum element in x
      a[1]: = a[n]; // places the left bottom most element at
                       the root
      RHeap(a, 1, n-1); // reheaps
}
```

Algorithm 4.11 Reheaps the elements

```
Algorithm RHeap (a, k, n)
// The two complete binary trees with roots 2k and 2k+1 are
// combined with node k resulting in a heap at k.
{
   i: = 2k;
   y: = a[k];
   while (i ≤ n) do
  // No node has a key greater than n and less than 1
   {
      if ((i < n) and (a[i] < a[i+1])) then i: = i+1;
  // compares the children
         if (y ≥ a[i]) then break;
  // location for y is found
```

```
            a[⌊i/2⌋]: = a[i];
            i:= 2*i;
        }
        a[⌊i/2⌋]: = y;
    }
```

Heaps are used to the greatest advantage to sort a given set of elements. Now we present the heap sort algorithm. First construct a heap with the given n elements. Delete its root (that contains the maximum value in the list), then reheap and again delete its root. In all, this process is repeated n times. We don't actually delete the root, but exchange the root with the node at the bottom of the heap. We observe that the following heap sort algorithm has $O(n \log n)$ run time as each insertion and deletion take $O(\log n)$ time. Algorithm 4.12 uses two functions: ReHeap (for readjusting the specified elements to form a Heap) and Adjust, which are given below.

Algorithm 4.12 Heapsort

```
Algorithm HeapSort(a, n)
// This algorithm sorts the elements of a[1..n]
   {
      CHeap(a,n) // construct a heap with the elements a[1 : n]
// interchange the root with the left bottom most element
         for j: = n to 2 step -1 do
            {
               // interchange a[j] with a[1]
            temp: = a[j];
            a[j]: = a[1];
            a[1]: = temp;
              RHeap(a, 1, j-1);
            }
   }
```

Algorithm 4.13 constructs a heap with n data items.

Algorithm 4.13 Constructs a heap

```
Algorithm CHeap (a,n)
// Used to create a heap with the elements from a[1 : n]
{
   for j: = ⌊n/2⌋ to 1 step -1 do
   RHeap(a, j, n);
}
```

We now give the actual heap sort algorithm.

Algorithm 4.14 A sorting algorithm

```
Algorithm Sort (a, n)
// Sorts the elements a[1 : n]
  {
     for j: = 1 to n do
        HeapInsert (a, j);
     for j: = n to 1 step -1 do
        {
           DeleteMax (a,n,x);
           a[j]: = x;
        }
  }
```

4.9 AVL TREES

We are interested in the structure of binary trees that lead to the implementation of dictionaries and priority queues with an insertion operation time of $O(\log_2 n)$ even in worst cases. One such tree is the AVL tree. We now consider this tree.

A quantity called "**balance factor**" (bf) is associated with each node of a binary tree. The balance factor of a node is defined as the difference between the heights of its left subtree and right subtree.

Example 4.2

Consider the binary tree:

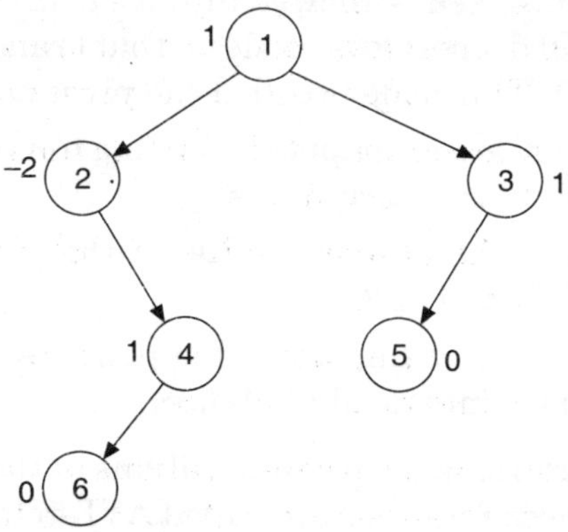

Figure 4.9 A binary tree.

The balance factors of nodes in the above binary tree are:

Node	Balance Factor (bf)
1	3 – 2 = 1
2	0 – 2 = –2
3	1 – 0 = 1
4	1 – 0 = 1
5	0 – 0 = 0
6	0 – 0 = 0

In Figure 4.9 balance factors of different nodes are shown by their side. In binary search trees, whenever insertions and deletions are made, 'balance' of nodes is disturbed. Any further search or insertion or deletion operations will take more time. All these times can be optimized by keeping the tree very nearly balanced all the time.

4.9.1 Height Balanced Trees

A binary tree is said to be height-balanced if all its nodes have a balance factor of either 1, 0 or –1. The method of 'balancing' a binary tree was introduced by two Russian mathematicians, viz. Adelson-Velkii and Landis in 1962 and the resulting binary trees are called AVL trees. The time of performing search or insertion or deletion operations is of order $O(\log_2 n)$ even in the worst cases.

Definition [AVL Trees]: An AVL tree is a binary tree in which: (1) the heights of the left and right subtrees of the root differ at most by 1 and (2) the left and right subtrees are again AVL trees.

4.9.2 AVL Techniques

The AVL technique is used to keep a tree in height balance, even after insertions. This technique is based on the following factors:

1. The node to be inserted is allowed to move to the appropriate branch, along the path of deepest level node on that branch that has a balance factor of 1 or –1. This node is called the pivot node.
2. All balance factors are recomputed, starting from the pivot node, along the insertion path as observed in step 1.
3. Observe whether the absolute value of the pivot node's balancing factor switched from 1 to 2.
4. If there is switch to manipulate tree pointers centred at the pivot node to bring back into height balance.

In step 4, we perform appropriate rotations of the subtree whose root is the pivot node. These rotations are called AVL-rotations. The following three figures illustrate AVL rebalancing, when a node is inserted in a height-balanced tree. In the following figures LT1, RT1, LT2 and RT2 denote, the left subtree of node 1, the right subtree of node 1, the left subtree of node 2 and the right subtree of node 2 respectively.

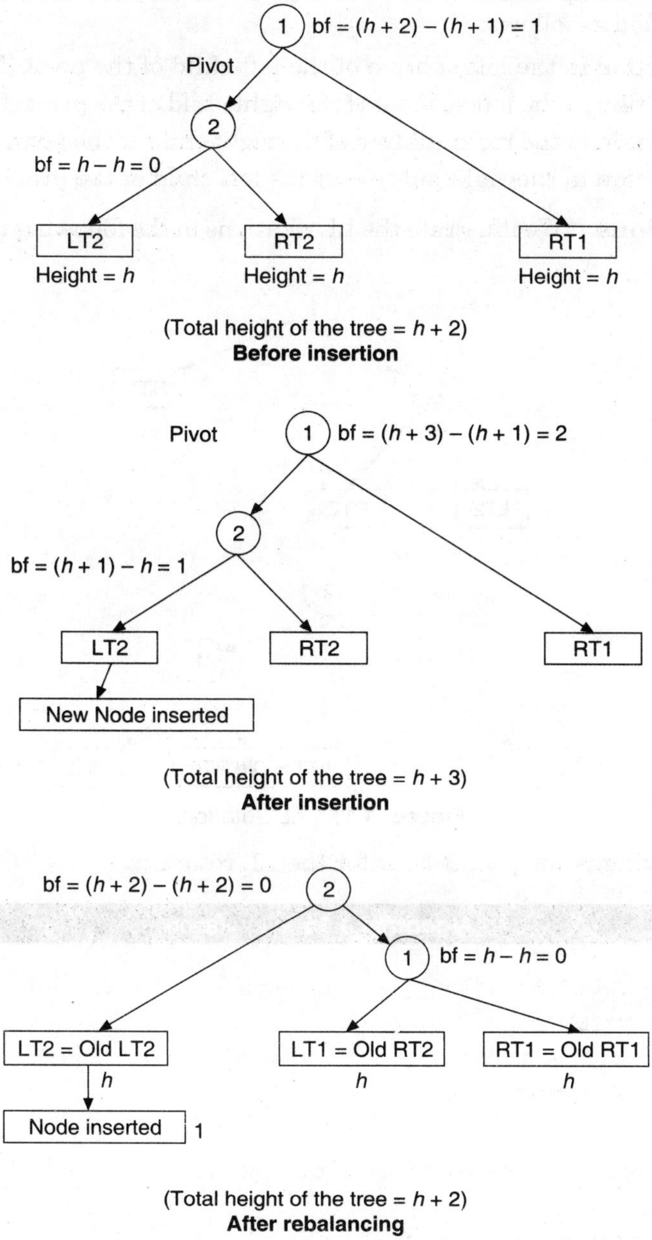

Figure 4.10 AVL tree-height balancing.

4.9.3 AVL Rotations

Based on the position at which an insertion takes place relative to the pivot node, manipulations, otherwise called rotations required to rebalance are classified as follows:

1. Insertion in the left subtree of the left child of the pivot (LL),
2. Insertion in the left subtree of the right child of the pivot (LR),
3. Insertion in the right subtree of the right child of the pivot (RR) and
4. Insertion in the right subtree of the left child of the pivot (LR).

LL Rotations: We illustrate the LL rotations in the following two figures.

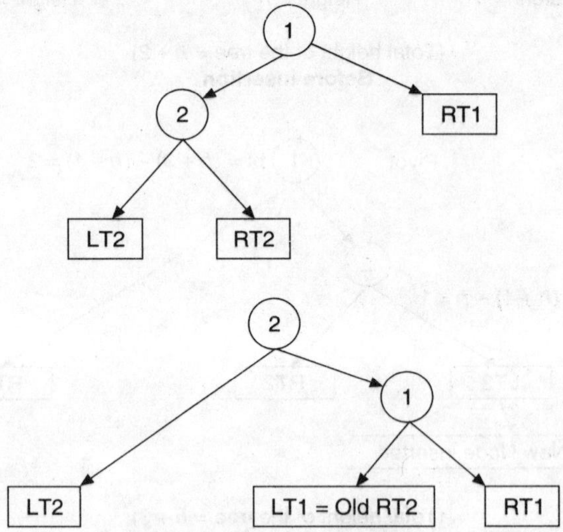

Figure 4.11 LL rotation.

Following is the pseudocode for the LL rotation:

Algorithm 4.15 LL rotation

```
P ← LLINK[PIVOT]
Q ← RLINK[P]
RLINK[P] ← PIVOT
LLLINK[PIVOT] ← Q
PIVOT ← P
//Readjust the balance factor (BF).
BF[Pivot] ← 0
BF[RLINK[PIVOT]] ← 0
return
```

RR ROTATIONS

The following figures illustrate the RR rotation:

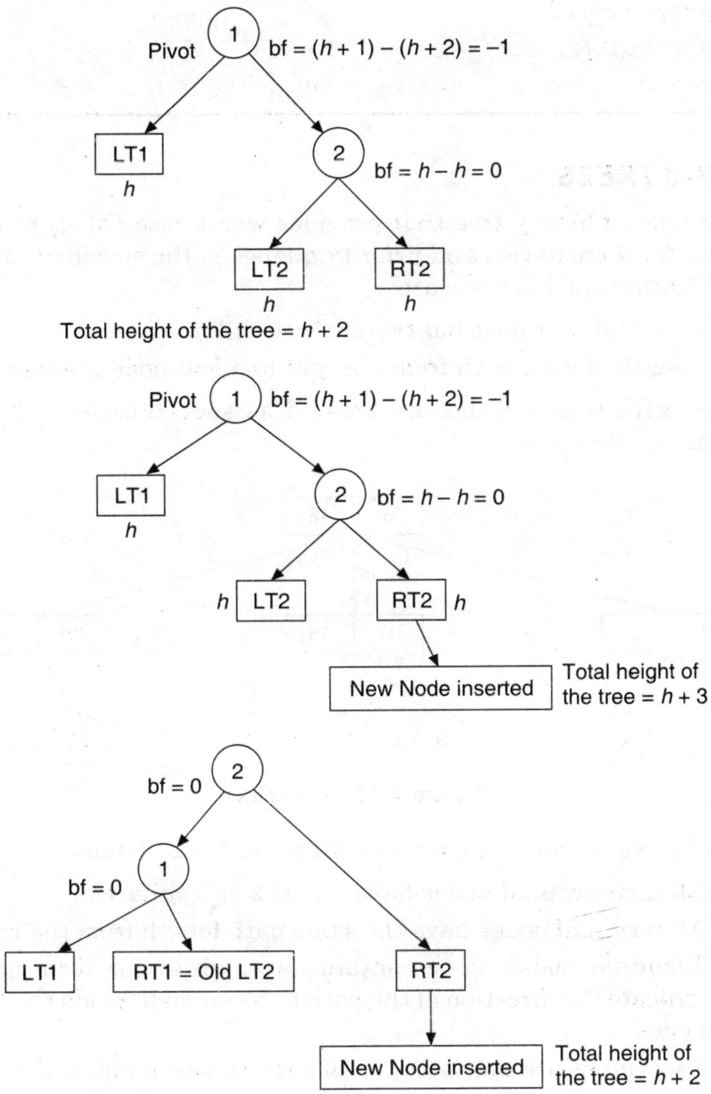

Figure 4.12 RR rotation.

Following is a pseudocode for RR rotation:

Algorithm 4.16 RR rotation

```
1. P ← RKINK[PIVOT]
2. Q ← LLINK[P]
3. LLINK[P] ← PIVOT
```

```
RLINK[Q] ← Q
PIVOT ← P
// Readjust the balance factor
BF[PIVOT] ← 0
BF[LLINK[PIVOT]] ← 0
return
```

4.10 2–3 TREES

Another type of binary tree that provides worst case $O(\log_2 n)$ time per operation for dictionaries and priority queues is the so called "2–3 tree". Marked features of 2–3 trees are:

1. Every interior node has two or three children.
2. Length of each path from the root to a leaf node is same.

Trees with 0 or 1 nodes are treated as special cases of 2–3 trees. Following is a 2–3 tree.

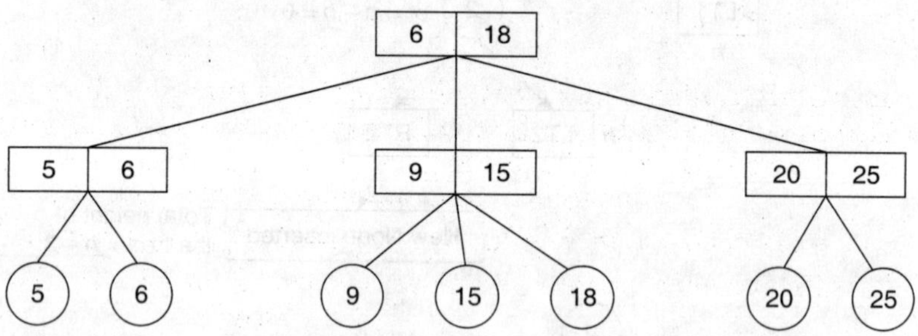

Figure 4.13 A 2–3 tree.

We observe the following points with respect to 2–3 trees:

1. All non-terminal nodes have either 2 or 3 children.
2. All terminal nodes have the same path length from the root node.
3. Terminal nodes alone contain data, while non-terminal nodes indicate the direction of the path to be taken to reach the terminal nodes.
4. Two values are associated with each non-terminal node.
5. While taking a path to reach a terminal node, the given key is compared with the two values as said in step 4. Then the following three cases will arise:
   (a) Key ≤ first value. Then the left branch is taken.
   (b) First value ≤ key ≤ Second value. Then the middle branch is taken.
   (c) Second value ≤ key. Then the right branch is taken.

4.10.1 Implementation of Membership Function

A record with key x is tested whether it is a member of the set of elements represented by a 2–3 tree as follows:

1. We move down the tree, comparing at a NODE, x with the value y corresponding to the least element descending from the second child of NODE.
2. If $x < y$, then we go for the first child of NODE.
3. If $x \geq y$, then
   (a) If NODE has only two children, we go for the second child of NODE.
   (b) If NODE has three children, then we compare x with z, the second value listed at NODE.
4. If $x < z$, then we move to the second child.
5. If $x \geq z$, we move to the third child.

Note that if at any stage, either $x = y$ or $x = z$, the search terminates.

4.10.2 Insertion

To insert x into a 2–3 tree, we proceed as follows:

1. We test membership of x in the set.
2. At the level just above of the terminal nodes, the children of any node p_x, should not include x. Otherwise insertion is not possible.
3. If insertion is possible, then the main cases that arise are:
   (a) Node p_x has only two children. Then make x the third child of node p_x by placing x at an appropriate level. For example, consider the 2–3 node.

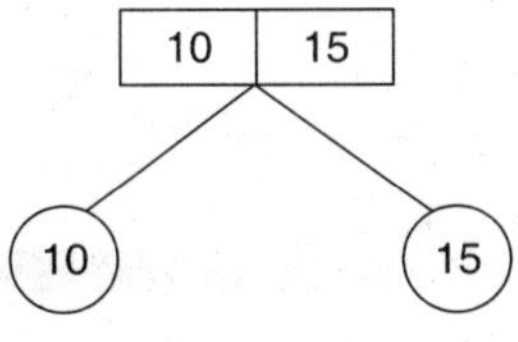

Figure 4.14 A 2–3 tree.

Then the following three 2–3 trees are generated depending upon $x = 8$ or 12 or 20.

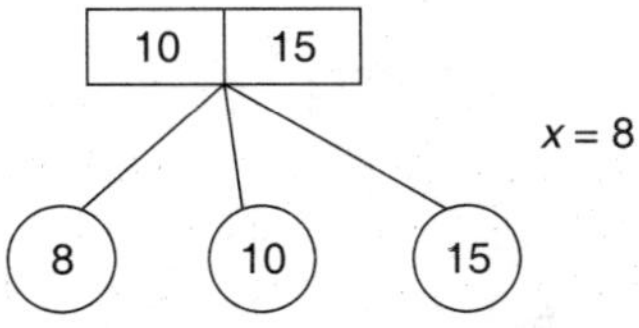

Figure 4.15 2–3 tree of Figure 4.14 after x = 8 is inserted.

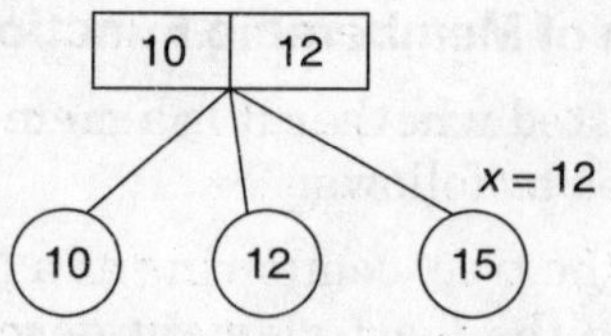

Figure 4.16 2–3 tree of Figure 4.14 after inserting the new node x = 12.

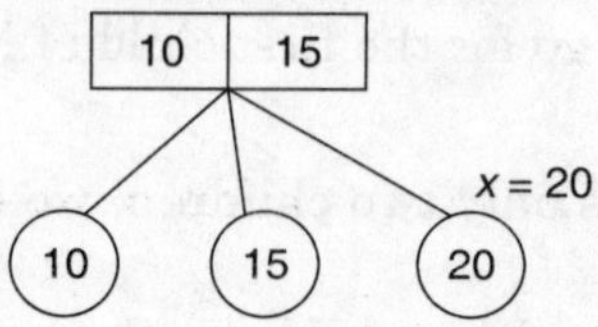

Figure 4.17 2–3 tree of Figure 4.14 after inserting the new node x = 20.

If we insert x = 25 in the 2–3 tree of Figure 4.17 of above, we obtain the trees

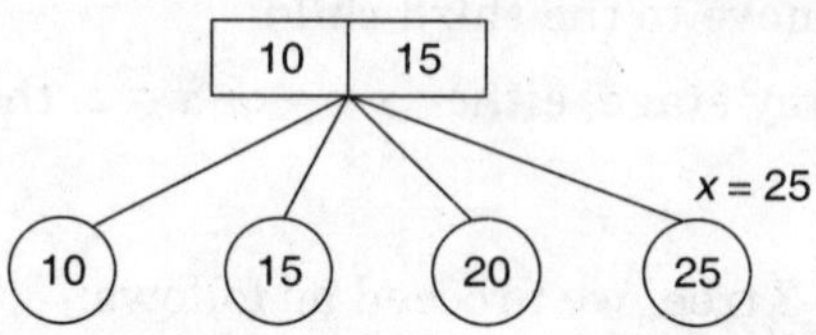

But any non-leaf node in a 2–3 tree can have 2 or 3 leaf nodes only. The above is split to obtain a 2–3 tree and the labels in the non-leaf nodes and the leaf nodes are shown as below:

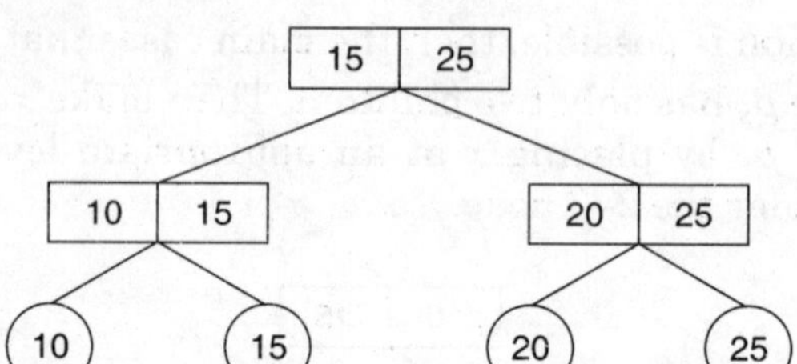

Example 4.3

Insert the element 9 into the following 2–3 tree.

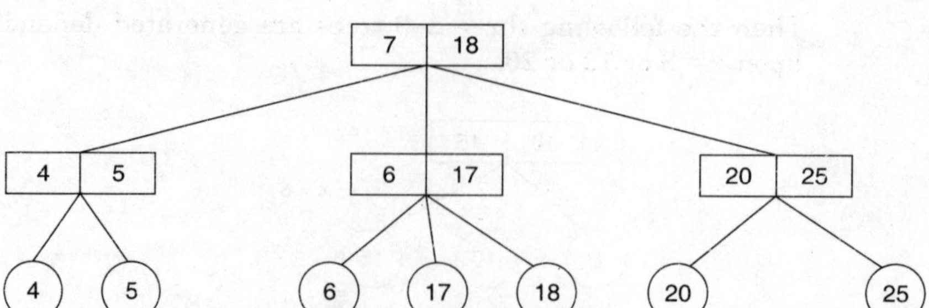

Figure 4.18 A 2–3 tree.

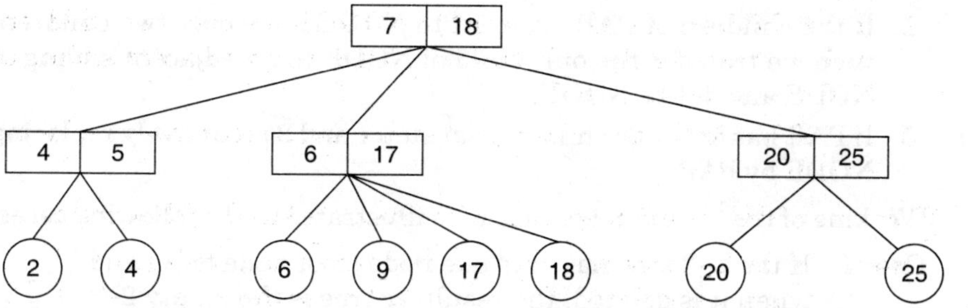

Figure 4.19 2–3 tree obtained immediately after inserting 9.

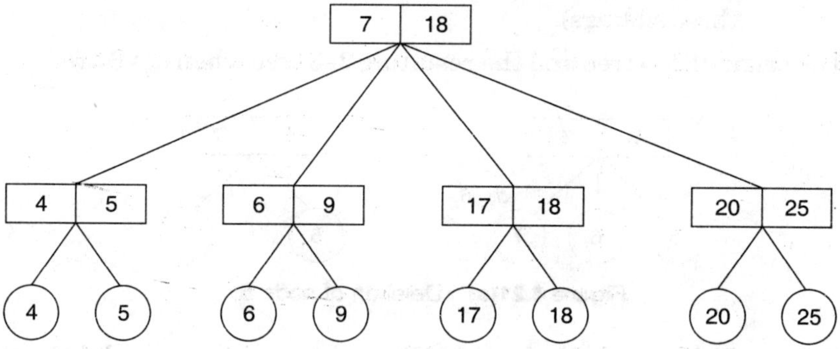

Figure 4.20(a) 2–3 tree during rebalancing.

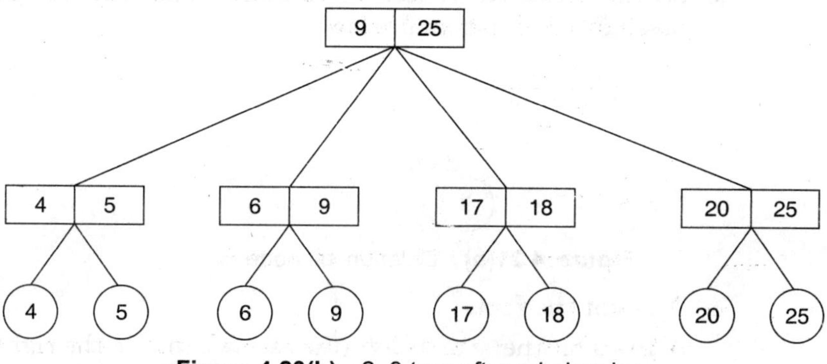

Figure 4.20(b) 2–3 tree after rebalancing.

4.10.3 Deletion

When the element to be deleted happens to be a leaf, its parent NODE will have only one child. If NODE happens to be the root, we delete the NODE and its only child becomes the new root. On the other hand, consider the parent PAR of NODE. The following three cases arise:

1. If PAR has another child, adjacent to NODE either to its left or right and PAR has three children, we transfer the appropriate child of these three to PAR.

2. If the children of PAR adjacent to NODE have only two children, then we transfer the only child of NODE to an adjacent sibling of NODE and delete NODE.
3. If PAR has only one child, repeat step 1 and 2 recursively replacing NODE by PAR.

Working of the "deletion operation" is illustrated in the following cases:

Case 1: If the 2–3 tree has only one node (root node only) and when it is deleted, the resulting tree is the empty 2–3 tree.

Case 2: (When the node to be deleted d is the child of its parent having three siblings).

The original 2–3 tree and the resulting 2–3 tree when $d = 6$ are:

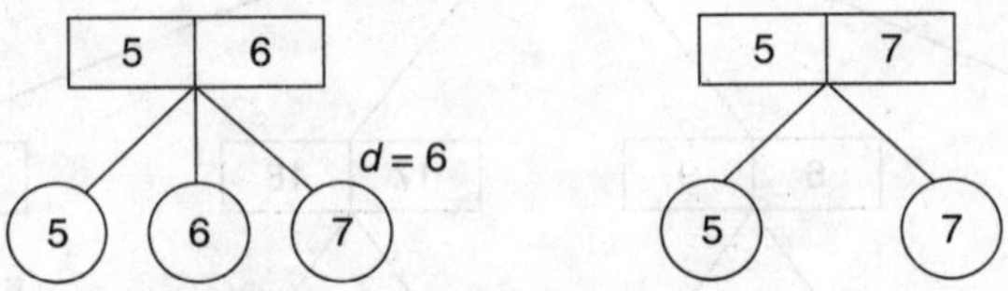

Figure 4.21(a) Deletion of node 6.

Case 3: If d is the child of a node N having two siblings c and d, then the following two possibilities arise:

(a) N is the root node. Delete N and d and make c the root. This possibility is illustrated below:

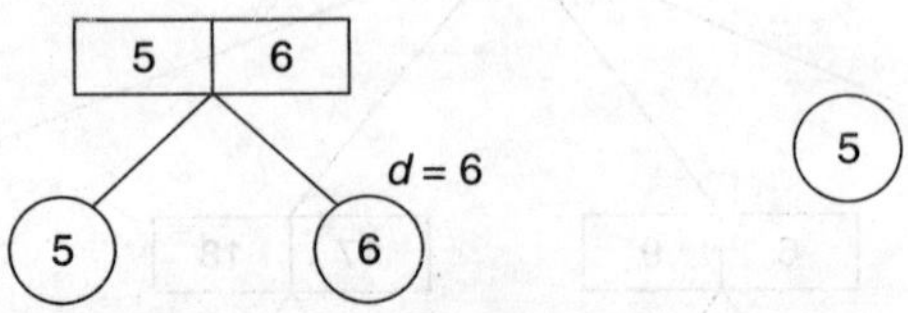

Figure 4.21(b) Deletion of node 6.

(b) N is not the root.

N has a brother e to its left (the case when is to the right of N is handled similarly) and e has two children only.

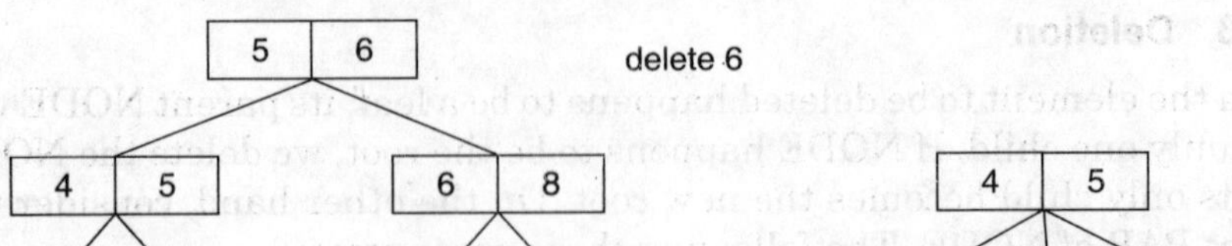

Figure 4.21(c) Deletion of node 6.

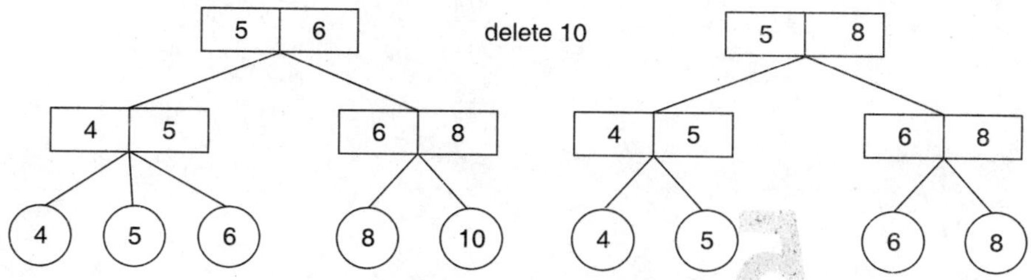

Figure 4.21(d) Deletion of node 10.

QUESTIONS

1. Explain the set operations: (1) MIN, (2) DELETE, (3) INSERT, (4) UNION, and (5) FIND.
2. Write a pseudocode for the implementation of UNION-FIND instruction with linked representation.
3. Given the three sets {1, 3, 8, 7}, {2, 4, 8} and {6} in which $n = 8$ with external names 1, 2, and 3 and with internal nodes 3, 2, and 1, give the data structure for UNION-FIND algorithm using a linked list.
4. What is a binary search tree? Write algorithms for the implementation of (1) Insertion, and (2) Deletion of a node in it.
5. Explain priority queues.
6. What are AVL trees? Explain AVL rotations with some examples.
7. What are 2–3 trees? Write algorithms for (1) Membership function, (2) Insertion, and (3) Deletion of a node in them.
8. What is a heap? Explain the technique of constructing a heap.
9. Write algorithms for (1) Insertion, and (2) Deletion of a node with respect to heaps.
10. Write an algorithm for combining two heaps.

CHAPTER 5

Dynamic Programming

5.1 INTRODUCTION

Dynamic programming is a powerful tool for designing algorithms. A better term to this is "recursive programming". This method is used recursively, wherein the final solution to a problem is obtained as a result of a sequence of decisions. A problem in n variables is divided into n sub problems each in one variable only.

In greedy method, only one sequence of decisions is made, whereas in dynamic programming many decision sequences may be generated, if necessary. Dynamic programming is based on the optimality principle due to Bellman.

5.2 PRINCIPLE OF OPTIMALITY

"The principle of optimality says that an optimal sequence of decisions has the property that whatever the initial state and decision taken there, the remaining decisions must constitute an optimal decision sequence with regard to the state resulting from the earlier decision".

Normally, this principle of optimality is expressed mathematically by a recurrence relation like

1. $f_0(x) = 0$,
   $f_i(x) = \text{Optimum of } \{f_{i-1}(x), f_{i-1}(x - w_i) + p_i\}$, for $i = 1$ to n.
2. $f_0(x) = 0$,
   $f_i(x) = \text{Optimum of } \{g_i(m_i)\, f_{i-1}(x - c_i m_i)\}$ for $i = 1$ to n.

Most of the problems such as 0/1 Knapsack optimal, multistage graph, optimal binary search tree problem, reliability design, the travelling salesman problem, flow shop scheduling problem and many other related problems can be solved very effectively using dynamic programming technique. We now consider dynamic programming formulations for these problems.

5.3 THE 0/1 KNAPSACK PROBLEM

The 0/1 knapsack problem differs from the knapsack problem considered earlier in the following sense. As before the maximum capacity of the knapsack is m and there are n items or objects. If an item i is selected then it is wholly put in the knapsack. The question of putting a fraction of this selected item does not arise. A solution to the 0/1 knapsack problem can be obtained by making a sequence of decisions on the variables $x_1, x_2, \ldots, x_n$ corresponding to the n objects. A decision on variable x_i involves determining which of the values 0 or 1 is to be assigned to it (whether to select or not to select the object i). Suppose that the decisions on x_i are made in the order $x_n, x_{n-1}, \ldots, x_1$. After a decision on x_n is made, we may be in one of two possible states: the capacity remaining in the knapsack is m and no profit has accrued or the capacity remaining is $m - w_n$ and a profit p_n has accrued. The remaining decisions to be made about $x_{n-1}, x_{n-2}, \ldots, x_1$ must be optimal with respect to the problem state resulting from the decision on x_n. Otherwise, $x_n, \ldots, x_1$ will not be optimal.

If $f_i(v)$ is the optimal solution to the KNAP(i + 1, n, v), then $f_0(m)$ is the required optimal solution to the KNAP(1, n, m) problem. The principle of optimality gives:

$$f_i(m) = \text{Max } \{f_{i+1}(m), f_{i+1}(m - w_{i+1}) + p_{i+1}\} \quad \text{for } i = 0, 1, 2, \ldots, n - 1.$$

We apply backward recursion for computing $f_0(m)$. We first compute f_{n-1}(m), then $f_{n-2}(m)$ and finally $f_0(m)$. The following example illustrates the working of this method. We note that f_n (anything) and also $f_p(q)$, for $p \geq q$ are not feasible and we set them to 0.

Example 5.1

Solve the 0/1 knapsack problem for $n = 3$, $(w_1, w_2, w_3) = (2, 4, 8)$, $(p_1, p_2, p_3) = (1, 3, 4)$ and $m = 10$.

Solution: The required solution is $f_0(10)$. From the above said recurrence relation, we have

$$\begin{aligned} f_0(10) &= \text{Max } \{f_1(10), f_1(10 - w_1) + p_1\} \\ &= \text{Max } \{f_1(10), f_1(8) + 1\} \end{aligned} \tag{5.1}$$

$$\begin{aligned} f_1(10) &= \text{Max } \{f_2(10), f_2(10 - w_2) + p_2\} \\ &= \text{Max } \{f_2(10), f_2(6) + 3\} \end{aligned} \tag{5.2}$$

Now

$$\begin{aligned} f_2(10) &= \text{Max } \{f_3(10), f_3(10 - w_3) + p_3\} \\ &= \text{Max } \{0, f_3(2) + 4\} \\ &= \text{Max } \{0, \underline{0 + 4}\} \\ &= 4 \end{aligned} \tag{5.3}$$

Now $\quad f_2(6) = \text{Max}\,\{f_3(6), f_3(6 - w_3) + p_3\}$

$= \text{Max}\,\{0, f_3(-2) + p_3\}$

$= \text{Max}\,\{0, -\infty\}$

$= 0 \qquad (5.4)$

Thus, $\quad f_1(10) = \text{Max}\,\{\underline{4}, 0 + 3\}$, from (2)

$= 4$

Also $\quad f_1(8) = \text{Max}\,\{f_2(8), f_2(8 - w_2) + p_2\}$

$= \text{Max}\,\{f_2(8), f_2(4) + 3\}$

$f_2(8) = \text{Max}\,\{f_3(8), f_3(8 - w_3) + p_3\}$

$= \text{Max}\,\{0, 0 + 4\}$

$= 4$

Now $\quad f_2(4) = \text{Max}\,\{f_3(4), f_3(4 - w_3) + p_3\}$

$= \text{Max}\,\{0, -\infty\}$

$= 0 \qquad (5.5)$

Thus, $\quad f_1(8) = \text{Max}\,\{4, 0 + 3\}$

$= 4 \qquad (5.6)$

Hence $\quad f_0(10) = \text{Max}\,\{4, \underline{4 + 1}\} = 5$

We conclude that the maximum profit is 5 units. By backtracking and noticing where from the maximum is coming in the computations of $f_0(10)$, $f_1(10)$, and $f_2(10)$, we note that the optimal solution is (1, 0, 1).

5.4 MULTISTAGE GRAPHS

Consider the following graph as shown in Figure 5.1.

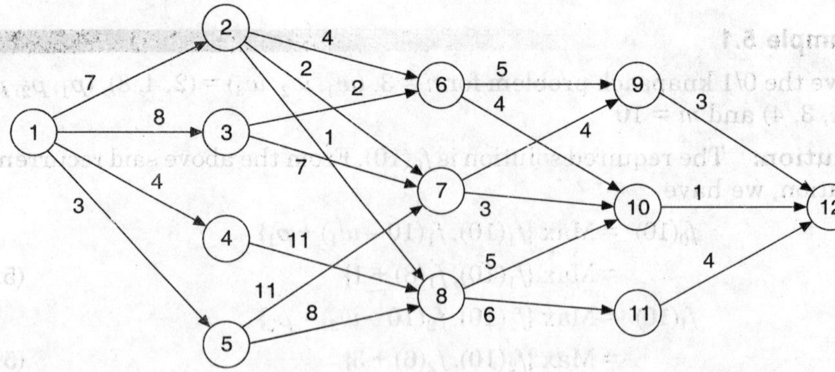

Figure 5.1 A multistage graph.

The problem here is a path of minimum weight to be obtained starting from node 1 and ending at node 12. We illustrate how dynamic programming technique can effectively be used in solving this problem.

We view this as a 5-stage problem. At each stage there are different states. The sets V_i of states at stage i are shown below:

$$V_1 = \{1\},$$
$$V_2 = \{2, 3, 4, 5\},$$
$$V_3 = \{6, 7, 8\},$$
$$V_4 = \{9, 10, 11\},$$
$$V_5 = \{12\}.$$

We write (i, j) as an edge in the graph such that $i \in V_k$ and $j \in V_{k+1}$ for some k. For instance, (3, 6) is an edge such that $3 \in V_2$ and $6 \in V_3$ and $c(3, 6) = 2$. $c(i, j)$ is the cost/weight/length of the edge (i, j). The cost from a node s to a node t is the sum of the costs of all the edges on the path.

A multistage graph $G = (V, E)$ is a directed graph in which V is partitioned into disjoint sub sets $V_1, V_2, \ldots, V_k$, $(k \geq 2)$. If (u, v) is an edge in E, then $u \in V_i$ and $v \in V_{i+1}$ for some i, $1 \leq i \leq k$.

The multistage graph problem is to find a minimum cost path from a given source node to a destination node.

Let $p(i, j)$ be a minimum cost path from node v_i to node v_j. Let cost (i, j) be the cost of this path. Then, the principle of optimality gives:

$$\text{cost}\,(i, j) = \min\,\{c(j, k) + \text{cost}\,(i + 1, k)\}$$
$$\text{for all } k \in V_{i+1} \text{ and } (j, k) \in E.$$

Note that

$$\text{cost}\,(k - 1, j) = c(j, t) \text{ if } (j, t) \in E \text{ and is set to a large value if } (j, t) \notin E.$$

Example 5.2

We have the following computations for the above 5-stage graph problem. One can apply either forward or backward recursion in solving these problems. We apply backward recursion.

For stage i = 5:

$$V_i = V_5 = \{12\}$$
$$\text{cost}\,(i, 12) = \text{cost}\,(5, 12) = 0.$$

For stage i = 4:

$V_i = V_4 = \{9, 10, 11\}$; $V_{i+1} = V_5 = \{12\}$

$\text{cost}\,(4, 9) = \text{Min}\{c(9, 12) + \text{cost}\,(5, 12)\} = \text{Min}\,\{3 + 0\} = 3$

$\text{cost}\,(4, 10) = \text{Min}\,\{c(10, 12) + \text{cost}\,(5, 12)\} = \text{Min}\,\{1 + 0\} = 1(*)$

$\text{cost}\,(4, 11) = \text{Min}\,\{c(11, 12) + \text{cost}\,(5, 12)\} = \text{Min}\,\{4 + 0\} = 4$

For stage i = 3:

$$V_i = V_3 = \{6, 7, 8\}$$
$$V_{i+1} = V_4 = \{9, 10, 11\}$$

$$\text{cost}\,(3, 6) = \text{Min}\,\{c(6, 9) + \text{cost}\,(4, 9), c(6, 10) + \text{cost}\,(4,10), c(6, 11) + \text{cost}\,(4, 11)\}$$

$$= \text{Min }\{5 + 3, 4 + 1, ...\} = 5$$

$$\text{cost }(3, 7) = \text{Min }\{c(7, 9) + \text{cost }(4, 9), c(7, 10) + \text{cost }(4, 10)$$
$$c(7, 11) + \text{cost }(4, 11)$$
$$= \text{Min }\{3 + 3, 2 + 1, ...\} = 3\ (*)$$

$$\text{cost }(3, 8) = \text{Min }\{c(8, 10) + \text{cost }(4, 10), c(8, 11) + \text{cost }(4, 11)\}$$
$$= \text{Min }\{4 + 1, 5 + 4\} = 5$$

For stage i = 2:

$$V_i = V_2 = \{2, 3, 4, 5\}$$
$$V_{i+1} = V_3 = \{6, 7, 8\}$$

$$\text{cost }(2, 2) = \text{Min }\{c(2, 6) + \text{cost }(3, 6), c(2, 7)$$
$$+ \text{cost }(3, 7), c(2,8) + \text{cost }(3, 8)\}$$
$$= \text{Min }\{3 + 5, 1 + 3, 1 + 5\} = 4\ (*)$$

$$\text{cost }(2, 3) = \text{Min }\{c(3, 6) + \text{cost }(3, 6), c(3, 7)$$
$$+ \text{cost }(3, 7), c(3, 8) + \text{cost }(3, 8)\}$$
$$= \text{Min }\{1 + 5, 6 + 3, ...\} = 6$$

$$\text{cost }(2, 4) = 15$$
$$\text{cost }(2, 5) = 12$$

For stage i = 1:

$$V_i = V_1 = \{1\},$$
$$V_{i+1} = V_2 = \{2, 3, 4, 5\}$$

$$\text{cost }(1, 1) = \text{Min }\{c(1, 2) + \text{cost }(2, 2), c(1, 3) + \text{cost }(2, 3), c(1, 4)$$
$$+ \text{cost }(2, 4), c(1, 5) + \text{cost }(2, 5)\}$$
$$= \text{Min }\{8 + 4, 6 + 6, 2 + 15, 1 + 12\} = 12$$

The cost of the minimum cost path is 16 units.
Backtracking, we obtain the path of minimum cost path as

$$12 \longrightarrow 10 \longrightarrow 7 \longrightarrow 2 \longrightarrow 1$$

That is

$$1 \longrightarrow 2 \longrightarrow 7 \longrightarrow 10 \longrightarrow 12$$

We now give the necessary algorithm for solving the multistage graph problem using dynamic programming approach—forward recursions.

Algorithm 5.1 Dynamic programming—Multistage graph-forward computation

```
Algorithm F_MultistageGraph(G, k, n, p)
// G = (V, E) is a multistage graph with n vertices.
// V is divided into k sub sets V1, V2, ..., Vk.
// Each Vi defines a stage (i).
```

```
// (u, v) is an edge in G implies that u ε V_r and v ε V_r+1 for
1 ≤ r ≤ n-1
// c[i, j] is the cost (weight) of edge (i, j).
// p[i : k]) is a minimum-cost path.
// totcost[j] stores the total minimum cost at stage j.
// d [i] stores the state selected at stage i.
{
   totcost[n]: = 0.0;
   for i: = n-1 to 1 step -1 do
   {
Let r be such that (i, r) be an edge of G and c[i, r]+totcost[r] is
minimum
      totcost[i]: = totcost[r] + c[i, r];
      d[i]: = r;
   }
      //Tracing the minimum-cost path.
      p[1]: = 1;
      p[k]: = n;
      for i: = 2 to k-1 do
         p[i]: = d[p[i-1]];
}
```

In a similar way, one can devise backward computation also.

5.5 ALL PAIRS SHORTEST PATHS

$G = (V, E)$ is a directed graph with n vertices, maintained in the memory as its cost adjacency matrix. Let cost $[i, j]$ be the cost of the edge (i, j).

Let cost $[i, i] = 0.0$. If there is no edge from i to j then let cost $[i, j]$ be set to a very large number. The all pairs shortest problem is to determine a matrix A such that $A[i, j]$ is the cost of the minimum-cost path from vertex i to the vertex j.

Let $A^k[i, j]$ denote the cost of the minimum-cost path from vertex i to vertex j going through no vertex of index greater than k.

$$A^k[i, j] = \text{Min}\,\{A^{k-1}[i, j], A^{k-1}[i, k] + A^{k-1}[k, j]\}, \quad \text{for } 1 \le k \le n \qquad (1)$$

Following is the concerned algorithm for the all-pairs shortest paths.

Algorithm 5.2 All pairs shortest paths

```
Algorithm AllPaths(cost, A, n)
// cost[1 : n, 1 : n ] is the cost adjacency matrix of a graph
//G = (V, E) with n vertices.
```

```
// A[p, q] is the cost of minimum-cost path from vertex p to
   vertex q.
// cost[p, p] = 0.0 for all p = 1 to n.
{
// Initialise the matrix A by setting A[i, j] = cost[i, j]
     for each i, j
for p := 1 to n do
   for q := 1 to n do
      A[p, q] := cost[p, q] ;
// Start computing the minimum-cost path matrix A
   for k := 1 to n do
      for p := 1 to n do
         for q := 1 to n do
            A[p, q] := min ( A[p, q] , A[p, k] + A[k, q] ) ;
}
```

Example 5.3

Apply the above algorithm and obtain all shortest paths of the directed graph

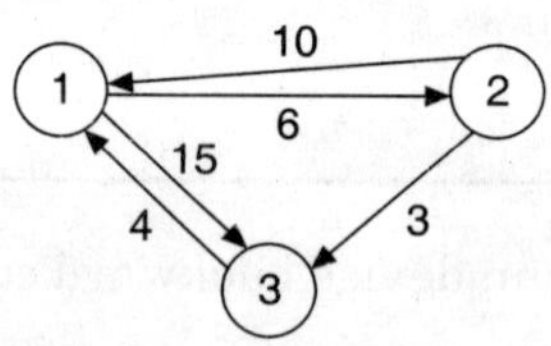

Here, $n = 3$. We, therefore, compute the three matrices $A^k(i, j)$ for $1 \le i, j \le 3$ using (1) as stated above, starting with the matrix.

A^0 = cost adjacency matrix of the given directed graph

$$A^0 = \begin{bmatrix} 0 & 6 & 15 \\ 10 & 0 & 3 \\ 4 & \infty & 0 \end{bmatrix}$$

<u>$k = 1$.</u>

$$A^1(1, 1) = \text{Min}\,\{A^0(1, 1) + A^0(1, 1), A^0(1, 1)\}$$
$$= \text{Min}\,\{0 + 0, 0\} = 0$$
$$A^1(1, 2) = \text{Min}\,\{A^0(1, 1) + A^0(2, 1), A^0(1, 2)\}$$
$$= \text{Min}\,\{0 + 6, 6\} = 6$$
$$A^1(1, 3) = \text{Min}\,\{A^0(1, 1) + A^0(1, 3), A^0(1, 3)\}$$
$$= \text{Min}\,\{0 + 5, 15\} = 15$$
$$A^1(2, 1) = \text{Min}\,\{A^0(2, 1) + A^0(1, 1), A^0(2, 1)\}$$
$$= \text{Min}\,\{10 + 0, 10\} = 10$$

$$A^1(2, 2) = \text{Min}\,\{A^0(2, 1) + A^0(1, 2), A^0(2, 2)\}$$
$$= \text{Min}\,\{10 + 6, 0\} = 0$$
$$A^1(2, 3) = \text{Min}\,\{A^0(2, 1) + A^0(1, 3), A^0(2, 3)\}$$
$$= \text{Min}\,\{10 + 15, 3\} = 3$$
$$A^1(3, 1) = \text{Min}\,\{A^0(3, 1) + A^0(1, 1), A^0(3, 1)\}$$
$$= \text{Min}\,\{4 + 0, 4\} = 4$$
$$A^1(3, 2) = \text{Min}\,\{A^0(3, 1) + A^0(1, 2), A^0(3, 2)\}$$
$$= \text{Min}\,\{4 + 6, \infty\} = 10$$
$$A^1(3, 3) = \text{Min}\,\{A^0(3, 1) + A^0(1, 1), A^0(3, 3)\}$$
$$= \text{Min}\,\{4 + 15, 0\} = 0$$

Thus,

$$A^1 = \begin{bmatrix} 0 & 6 & 15 \\ 10 & 0 & 3 \\ 4 & 10 & 0 \end{bmatrix}$$

<u>$k = 2$.</u>

$$A^2(1, 1) = \text{Min}\,\{A^1(1, 2) + A^1(2, 1), A^1(1, 1)\}$$
$$= \text{Min}\,\{6 + 10, 0\} = 0$$
$$A^2(1, 2) = \text{Min}\,\{A^1(1, 2) + A^1(2, 2), A^1(1, 2)\}$$
$$= \text{Min}\,\{6 + 0, 6\} = 6$$
$$A^2(1, 3) = \text{Min}\,\{A^1(1, 2) + A^1(2, 3), A^1(1, 3)\}$$
$$= \text{Min}\,\{6 + 3, 15\} = \{9, 15\} = 9$$

Thus,

$$A^2 = \begin{bmatrix} 0 & 6 & 9 \\ 10 & 0 & 3 \\ 4 & 10 & 0 \end{bmatrix}$$

<u>$k = 3$.</u>

$$A^3(1, 1) = \text{Min}\,\{A^2(1, 3) + A^2(3, 1), A^2(1, 1)\}$$
$$= \text{Min}\,\{9 + 4, 0\} = 0$$
$$A^3(1, 2) = \text{Min}\,\{A^2(1, 3) + A^2(3, 2), A^2(1, 2)\}$$
$$= \text{Min}\,\{9 + 10, 6\} = 6$$
$$A^3(1, 3) = \text{Min}\,\{A^2(1, 3) + A^2(3, 3), A^2(1, 3)\}$$
$$= \text{Min}\,\{9 + 0, 9\} = 9$$
$$A^3(2, 1) = \text{Min}\,\{A^2(2, 3) + A^2(3, 1), A^2(2, 1)\}$$
$$= \text{Min}\,\{3 + 4, 10\} = 7$$
$$A^3(2, 2) = \text{Min}\,\{A^2(2, 3) + A^2(3, 2), A^2(2, 2)\}$$
$$= \text{Min}\,\{9 + 10, 0\} = 0$$

$$A^3(2, 3) = \text{Min}\,\{A^2(2, 3) + A^2(3, 3), A^2(2, 3)\}$$
$$= \text{Min}\,\{3 + 0, 3\} = 3$$
$$A^3(3, 1) = \text{Min}\,\{A^2(3, 3) + A^2(3, 1), A^2(3, 1)\}$$
$$= \text{Min}\,\{0 + 4, 4\} = 4$$
$$A^3(3, 2) = \text{Min}\,\{A^2(3, 3) + A^2(3, 2), A^2(3, 2)\}$$
$$= \text{Min}\,\{0 + 10, 10\} = 10$$
$$A^3(3, 3) = \text{Min}\,\{A^2(3, 3) + A^2(3, 3), A^2(3, 3)\}$$
$$= \text{Min}\,\{0 + 0, 0\} = 0$$

Thus,

$$A^3 = \begin{bmatrix} 0 & 6 & 9 \\ 7 & 0 & 3 \\ 4 & 10 & 3 \end{bmatrix}$$

This matrix gives the shortest paths from any node to any node.

5.6 RELIABILITY DESIGN

We now apply dynamic programming technique in designing a unit consisting of several devices connected in series. Let r_i be the reliability (performance measure) of the device D_i. If required the devices are duplicated to improve reliability by connecting multiple copies of the same device in parallel. In the configuration of Figure 5.2, for example, at stage I three copies of device D_1 are connected.

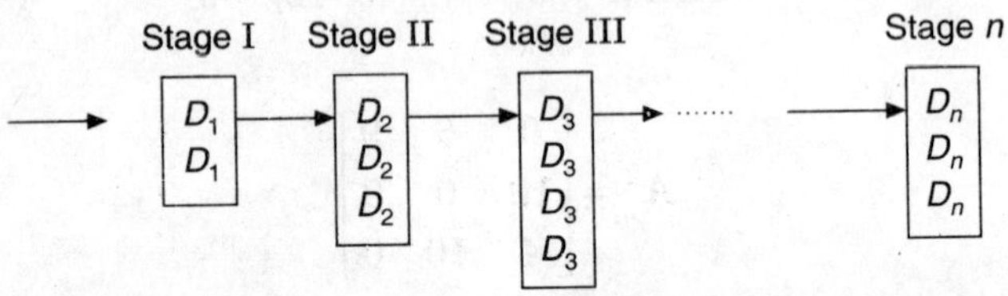

Figure 5.2 Multiple devices connected in Parallel in each stage.

Let there be n copies of the device D_i at stage i. Then the probability that all n_i will not function well is $(1 - r_i)^{ni}$. Hence, reliability of the system at stage i is $1 - (1 - r_i)^{n_i}$.

Let us assume that the reliability of stage i is a function like $\varphi(n_i)$. The reliability of the system is $\Pi\ \varphi(n_i)$ for $1 \le i \le n$.

The problem is to decide the optimal number of devices that are to be used at stage i. Let c_i be the cost of each unit of device i and let c be the maximum allowable cost of the system. Then the mathematical model of this reliability problem is:

Maximize $\quad \Pi\ \varphi(n_k)$

Subject to $\quad \Sigma c_k n_k \le c$,

for $1 \le k \le n$, and n_k is an integer greater than or equal to 1.

Dynamic programming technique is applied in solving this problem. Since we can assume each $c_k > 0$, each m_i must be in the range of $1 \le n_k \le u_k$, where

$$u_k = \lfloor (c + c_k - \Sigma c_k)/c_k \rfloor \quad \text{for } k = 1, 2, \ldots, n.$$

An optimal solution $n_1, n_2, \ldots, n_n$ is the result of a sequence of decisions, one decision for each m_i. Let $f_i(x)$ represent the maximum value of $\Pi\varphi(n_i)$ subject to the constraints $\Sigma c_j n_j \le x$ and $1 \le n_j \le u_j$, $1 \le j \le i$. The last decision made requires one to choose from $\{1, 2, \ldots, u_n\}$. Once a decision has been made for m_n, the remaining decisions must be such as to use the remaining funds $c - c_n n_n$ in an optimal way. The principle of optimality holds and

$$f_n(c) = \max \{\varphi_n(n_n) f_{n-1}(x - c_i n_i)\}$$

For any $f_i(x)$, $i \ge 1$, this equation generalizes to

$$f(x) = \max \{\varphi(n_i)\, f_{i-1}(x - c_i n_i)\} \quad \text{for } 1 \le n_i \le u_i$$

Initially $f_0(x) = 1$ for all x, $0 \le x \le c$.

We solve the above recurrence equations using the method similar to that used for the knapsack problem. Let S^i consist of tuples of the form (f, x), where $f = f_i(x)$. The dominance rule (f_1, x_1) dominates (f_2, x_2) if and only if $f_1 \ge f_2$ and $x_1 \le x_2$. Dominated tuples will be discarded.

Example 5.4

Consider the problem of designing a three-stage system with the following data. The costs are ₹ 45, ₹ 25 and ₹ 30 respectively. The cost of the system cannot be more than ₹ 150. The reliability of the devices are .8, .75 and .6 respectively.

Here,

$$c_1 = 45, \qquad c_2 = 25, \qquad c_3 = 30$$
$$r_1 = 0.8, \qquad r_2 = 0.75, \qquad r_3 = 0.6$$
$$c = 150.$$

$uk_i = \lfloor (c + c_k - \Sigma c_k)/c_k \rfloor$ gives

$$u_1 = \lfloor (150 + 45 - (45 + 25 + 30))/45 \rfloor = \lfloor 2.22 \rfloor = 2$$

Similar computations give: $u_2 = 3$, $u_3 = 2$.

We begin by setting $S^0 = \{(1, 0)\}$.

Since, $1 \le n_1 \le u_1$, we have, $1 \le n_1 \le 2$ and this gives $n_1 = 1, 2$.

Therefore, $j = 1$ or 2.

For $n_1 = 1$

$$\varphi_1(n_1) = \varphi_1(1) = 1 - (1 - r_1)^{n_1} = 1 - (1 - 0.8)^1 = 0.8$$

$$j{*}c_1 = 1{*}45 = 45$$

$$S_1^1 = S^0 * \{(\varphi_1(n_1), j{*}c_1)\} = \{(1, 0)\} * \{(0.8, 45)\} = \{(1 * 0.8, 0 + 45)\}$$
$$= \{(0.8, 45)\}$$

For $n_1 = 2$

$i = 1, j = 2$. This implies $n_i = n_1 = j = 2$

$$\varphi_i(n_i) = \varphi_1(n_1) = \varphi_1(2) = 1 - (1 - r_1)^{n_1} = 1 - (1 - .8)^2 = 0.96$$

$$j * c_i = 2 * c_1 = 2 * 45 = 90$$

$$S_2^1 = S^0 * \{(\varphi_1(n_1), j * c_1)\}$$

$$= \{(1, 0)\} * \{(\varphi_1(2), 2 * 45)\}$$

$$= \{(1, 0)\} * \{(0.96, 90)\}$$

$$= \{(0.96, 90)\}$$

$S^1 = S_1^1 + S_2^1$ Here, it is the merge operation.

$$= \{(0.8, 45), (0.96, 90)\}.$$

Note that we calculated S_1^1 and S_2^1, because $u_1 = 2$ and as such $j = 1$ or 2 ($j \le u_1$).

We now compute $S_j^2 = S_j^i$, for $j = 1, 2, 3$ as $j \le u_2 = 3$

$S_1^2 : i = 2, j = 1 = n_i = n_2$

$$\varphi_i(n_i) = \varphi_2(n_2) = \varphi_2(1) = 1 - (1 - r_2)^{n_1} = 1 - (1 - 0.75)^1 = 0.75$$

$$j * c_i = 1 * c_2 = 1 * 25 = 25$$

$$S_1^2 = S^1 * \{(\varphi_i(n_i), j * c_i)\}$$

$$= \{(0.8, 45), (0.96, 90)\} * \{(0.75, 25)\}$$

$$= \{(0.8 * 0.75, 45 + 25), (0.96 * 0.75, 90 + 25)\}$$

$$= \{(0.6, 70), (0.72, 115)\}$$

$S_2^2 : i = 2, j = 2, c_i = c_2 = 25$

$$\varphi_i(n_i) = \varphi_2(n_2) = \varphi_2(2) = 1 - (1 - r_2)^{n_2} = 1 - (1 - 0.75)^2 = 0.9375$$

$j * c_i = 2 * c_2 = 2 * 25 = 50$ is added.

$$S_2^2 = S^1 * \{(\varphi_i(n_i), j * c_i)\} = S^1 * \{\varphi_2(n_2), 50)\}$$

$$= \{(0.8, 45), (0.96, 90)\} * \{(0.9375, 50)\}$$

$$= \{(0.8 * 0.9375, 45 + 50), (0.96 * 0.9375, 90 + 50)\}$$

$$= \{(0.75, 95), (0.9, 140)\}$$

$S_3^2 : i = 2, j = 3 = n_i = n_2$

$$\varphi_i(n_i) = \varphi_2(n_2) = \varphi_2(3) = 1 - (1 - r_i)^{n_2}$$

$$= 1 - (1 - r_2)^{n_2} = 1 - (1 - 0.75)^3 = 0.984375$$

$j * c_i = 3 * c_2 = 3 * 25 = 75$ is added.

$$S_3^2 = S^1 * \{ (\varphi_i(n_i), j * c_i)\}$$

$$= S^1 * \{(\varphi_2(n_2), 75)\} = S^1 * \{(\varphi_2(3), 75)\}$$

$$= \{(0.8, 45), (0.96, 90)\} * \{ (0.984375, 75)\}$$

$$= \{(0.8 * 0.984375, 45 + 75), (0.96 * 0.984375, 90 + 75)\}$$

$$= \{ (0.7875, 120), (0.945, 165)\}$$

Discard the pair (.945, 165) as the cost of the system cannot be more than ₹ 150/- and take S_3^2 as

$$S_3^2 = \{(0.7875, 120)\}$$

$$S^2 = S_1^2 + S_2^2 + S_3^2 \quad \text{Here, + means 'merging'}$$

$$= \{(0.6, 70), (0.72, 115), (0.75, 95), (0.9, 140), (0.7875, 120)\}$$

We apply the purge rule or the rule of dominance to (.72, 115) and (.75, 95) and discard the pair (0.72, 115) pair from S^2 and take

$$S^2 = \{(0.6, 70), (0.75, 95), (0.9, 140), (0.7875, 120)\}$$

We now compute $S_j^3 = S_1^3$, for $j \le u_i = u_3 = 2$, that is for $j = 1, 2$.
We finally compute $S^3 = S_1^3 + S_2^3$.
For computing S_1^3, we identify that $i = 3, j = 1 = n_i = n_3$

$$\varphi_i(n_i) = \varphi_3(n_3) = \varphi_3(1) = 1 - (1 - r_3)^1 = .6$$

$$j * c_i = 1 * c_3 = 30$$

$$S_1^3 = S^2 * \{(\varphi_i(n_i), j * c_i)$$

$$= \{(.6, 70), (.75, 95), (.9, 140), (.7875, 120), (.945, 165)\} * (.6, 30)\}$$

$$= \{(.36, 100), (.45, 125), (.54, 170), (.47, 150), (.567, 195)\}$$

For the above said reasons we take

$$S_1^3 = \{(.36, 100), (.45, 95), (.9, 140), (.47, 150)\}$$

We now compute S_2^3.
For this computation, we observe that $i = 3, j = 2 = n_i = n_3$

$$\varphi_i(n_i) = \varphi_3(n_3) = \varphi_3(2) = 1 - (1 - r_3)^2 = 1 - (1 - .6)^2 = .84$$

$$j * c_i = 2 * 30 = 60$$

$$S_2^3 = S^2 * \{(\varphi_i(n_i), j * c_i)\}$$

$$= \{(0.6, 70), (.75, 95), (.9, 140), (.7875, 120)\} * \{(.84, 60)\}$$

$$= \{(.50, 130)\}, \text{ by considering the feasibility aspect.}$$

Now, $\quad S^3 = S_1^3 + S_2^3 \quad$ (Merging)

$$S^3 = \{(.36, 100), (0.45, 95), (.9, 140), (.47, 150)\} + \{(.50, 130)\} \quad \text{(Merging)}$$

$$= \{(0.36, 100), (0.45, 95), (0.9, 140), (.47, 150), (0.50, 130)\}$$

$$= \{(.45, 95), (.9, 140, (.47, 150), (.5, 130)\}$$

by applying the purging rule.

Presence of the pair (.47, 150) in S^3 indicates that a maximum reliability of 0.47 can be achieved at a cost of ₹ 150/-. Thus the best design is $n_1 = 1$, $n_2 = 3$, and $n_3 = 1$.

5.7 THE TRAVELLING SALESPERSON PROBLEM

This is a famous problem that can be solved by applying dynamic programming technique. Let $G = (V, E)$ be a directed graph that is represented by its edge cost adjacency matrix. We take the cost of the edge (i, j) to be c_{ij} (> 0). We set $c_{ij} = \infty$ if there is no edge from the vertex i to the vertex j. Let the number of vertices (= $|V|$) be n (for some integer $n > 1$). The travelling sales person problem (TSP problem) is to start from one vertex and then visit each and every other vertex once and only one and come back to the starting vertex so the sum of costs of visits to all these vertices is minimum. In other words, the travelling salesperson problem is to find a **tour** in G of minimum cost.

Without any loss of generality, we imagine our tour to start from vertex 1 and ends at the vertex 1. It is simple to understand that every tour consists of an edge $(1, k)$ for some $k \varepsilon V - \{1\}$ and a path from vertex k to 1. The path from vertex k to vertex 1 goes through each vertex in $V - \{1, k\}$ exactly once. We speak of the optimal path. Let $ls(i, S)$ be the length of a shortest path starting at vertex i, going through all vertices in a set S and terminating at vertex 1. The function $ls(1, V - \{1\})$ is the length of an optimal salesperson tour is given by $ls(1, V - \{1\})$.

The principle of optimality gives

$$ls(i, S) = \min \{c_{ij} + ls(j, S - \{j\})\}, \quad \text{where } i \notin S \text{ and } j \in S$$

That is

$$ls(1, V - \{1\}) = \min \{c_{1k} + ls(k, V - \{1, k\})\}, \quad \text{where } 2 \le k \le n$$

Using the above optimality principle, we solve the TSP problem recursively. Initially, we set $ls(i, \varphi) = c_{i1}$ for all $i = 2, 3, \ldots, n$. This method of solution is explained through the following problem.

Example 5.5

Solve the travelling salesperson problem for the cost-adjacency matrix is:

$$\begin{bmatrix} 0 & 10 & 15 & 20 \\ 5 & 0 & 9 & 10 \\ 6 & 13 & 0 & 12 \\ 8 & 8 & 9 & 0 \end{bmatrix}$$

To begin with, we have

$$ls(2, \varphi) = c_{21} = 5$$
$$ls(3, \varphi) = c_{31} = 6$$
$$ls(4, \varphi) = c_{41} = 8$$

For $|S| > 1$, we use the formula

$$ls(i, S) = \min \{c_{ij} + ls(j, S - \{j\})\} \text{ minimum over } j \in S$$

Computations for $|S| = 2$

$$ls(2, \{3\}) = c_{23} + ls(3, \varphi) = 9 + 6 = 15$$
$$ls(2, \{4\}) = c_{24} + ls(4, \varphi) = 10 + 8 = 18$$
$$ls(3, \{2\}) = c_{32} + ls(2, \varphi) = 13 + 5 = 18$$
$$ls(3, \{4\}) = c_{34} + ls(4, \varphi) = 12 + 8 = 20$$
$$ls(4, \{2\}) = c_{42} + ls(2, \varphi) = 8 + 5 = 13$$
$$ls(4, \{3\}) = c_{43} + ls(3, \varphi) = 9 + 6 = 15$$

Computations for $|S| = 3$

$$ls(2, \{3, 4\}) = \min \{c_{23} + ls(3, \{4\}), c_{24} + ls(4, \{3\})\}$$
$$= \min \{9 + 20, 10 + 15\} = 25$$
$$ls(3, \{2, 4\}) = \min \{c_{32} + ls(2, \{4\}), c_{34} + ls(4, \{2\})\}$$
$$= \min \{13 + 18, 12 + 13\} = 25$$
$$ls(4, \{2, 3\}) = \min \{c_{42} + ls(2, \{3\}, c_{43} + ls(3, \{2\})\}$$
$$= \min \{8 + 15, 9 + 16\} = 23$$

Finally, for $|S| = 4$,

$$ls(1, \{2, 3, 4\}) = \min \{c_{12} + ls(2, \{3, 4\}), c_{13} + ls(3, \{2, 4\}), c_{14} + ls(4, \{2, 3\})\}$$
$$= \min \{10 + 25, 15 + 25, 20 + 23\} = 35$$

Hence, **we observe that an optimal tour of the given graph has length 35 units.**

The actual tour can be constructed using backtracking.

The actual minimum-length tour is constructed by retaining with each $ls(i, S)$ the value of j that minimizes the right-hand side of $ls(i, S)$ in the equation

$$ls(i, S) = \min \{c_{ij} + ls(j, S - \{j\})\}, \text{ minimum taken over } j \in S$$

Let $J(i, S)$ be this value.

Then $$J(1, \{2, 3, 4\}) = 2.$$

Thus the tour starts from 1 and goes to 2.

Consider $ls(2, \{3, 4\})$, that contributed the minimum in the previous step.

$$J(2, \{3, 4\}) = 4$$

Thus, the next edge is (2, 4).

Now, consider $ls(4, \{3\})$.

$$J(4, \{3\}) = 3.$$

The next edge is (4, 3).

The optimal tour is: $1 \to 2 \to 4 \to 3 \to 1$.

5.8 A SCHEDULING PROBLEM

Consider that there are n jobs (programs). Each job requires m independent tasks (sub-programs) $T_{1i}, T_{2i}, \ldots, T_{mi}$ for $1 \le i \le n$ to be executed. Task T_{ji} is to be executed on processor P_j for $1 \le j \le m$. Suppose that the time required to complete task T_{ji} is t_{ji}. By a schedule for the n jobs, we mean an assignment of tasks to time intervals on the processors. We suppose that any processor can handle one and only one job at any time slot. To ensure continuous processing, we assume that for any job i the processing of task $T_{ji}, j > 1$, cannot be started until task T_{j-1i} has been completed. Consider the following example, wherein there are two jobs and three processors, the task times are given by the matrix:

$$A = \begin{bmatrix} 2 & 1 \\ 4 & 3 \\ 3 & 4 \end{bmatrix} = \begin{bmatrix} T_{11} & T_{12} \\ T_{21} & T_{22} \\ T_{31} & T_{32} \end{bmatrix}$$

One possible schedule for this problem is:

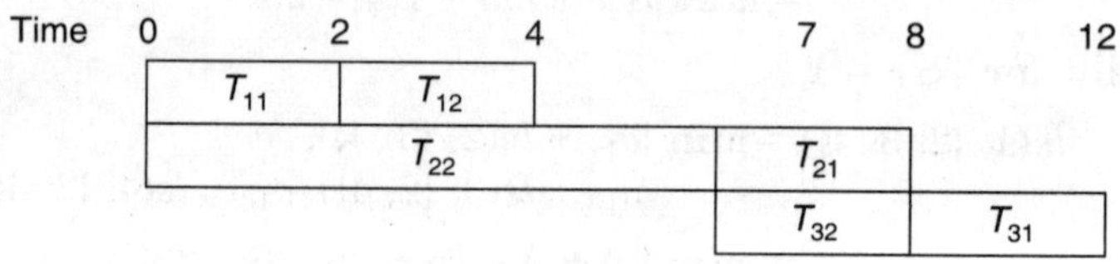

Figure 5.3 A possible schedule.

$f_i(S)$ denotes the finish time of the ith job at which all tasks of that job are completed in a time schedule S. From the above figure, we observe that $f_1(S) = 14$ and $f_2(S) = 10$. The finish time of a schedule is then

$$F(S) = \max \{f_i(S)\}, \qquad \text{for } 1 \le i \le n$$

The mean finish time MFT(S) is defined to be

$$\text{MFT}(S) = \frac{1}{n} \sum f_i(s), \qquad \text{for } 1 \le i \le n$$

An optimal finish time (OFT) schedule for a given set of jobs is a non-primitive schedule (this is a type of schedule in which processing of a task on any processor does not stop until the task is completed) S for which $F(S)$ is minimum over all non-primitive schedules S. A preemptive optimal finish time (POFT) schedule, optimal mean finish time (OMFT) schedule, and preemptive optimal mean finish time (POMFT) schedule are defined in a similar way. To make the notation easy, we denote hereafter t_{1i} and t_{2i} by a_i and b_i respectively. The different schedules for the n jobs are the different permutations of $(1, 2, 3, \ldots, n)$.

The optimal schedule is obtained by following the two steps given below:

1. We take the single array containing a_i's and b_j's and sort them in non-decreasing order.

2. Let at stage j, the number in the sequence be a_j. If job j has not yet been scheduled, schedule job j at the leftmost available spot. On the other hand, if the number is b_j and job j has not yet been scheduled, schedule job j at the rightmost available spot. We go to the next number if j has already been scheduled.

We illustrate this method in the following example.

Example 5.6

Let $n = 4$, $(a_1, a_2, a_3, a_4) = (2, 4, 10, 12)$ and $(b_1, b_2, b_3, b_4) = (5, 1, 11, 18)$.

The sorted sequencing of a's and b's is $(b_2, a_1, a_2, b_1, a_3, b_3, a_4, b_4) = (1, 2, 4, 5, 10, 11, 12, 18)$. Let os_1, os_2, os_3, and os_4 be the optimal schedule. Since the smallest number is b_2, we get $os_4 = 2$. The next number is a_1 and we get $os_1 = 3$. The next smallest number is a_2. Job 2 has already been scheduled. The next number is b_1. Job 1 has already been scheduled. The next number is a_3 and we get os_3. This leaves σ_3 free and job 4 unscheduled. Thus $os_3 = 4$. Hence the optimal sequence: (3, 1, 4, 2).

5.9 OPTIMUM BINARY SEARCH TREE (OBST) PROBLEM

We apply dynamic programming technique to solve the OBST problem. Let T be a binary tree with n nodes with an ordered set of keys $K = \{k_1, \ldots, k_n\}$. That is, key of node n_i is k_i and $k_1 < k_2 \ldots < k_n$. We construct a binary search tree with these keys. Let p_i be the probability that the search is for k_i. It may be possible that a search is for a value not in K. Such keys are called "dummy keys". We consider the ordered $n + 1$ dummy keys: $d_0, d_1, \ldots, d_n$, in such a manner that $d_0 \leq k_i$, $d_n \geq k_i$ for $1 \leq i \leq n$, and d_i represents all values between k_i and k_{i+1} for $1 < i < n$. Let q_i be the probability for d_i.

Consider a subtree of the given BST with keys $k_i, \ldots, k_j$, for $1 \leq i \leq j \leq n$ and with leaf nodes having the dummy keys $d_{i-1}, \ldots, d_j$. The principle of optimality ensures us that if an optimal binary search tree T has a subtree with keys $k_i, \ldots, k_j$, then this subtree must also be optimal for the subproblem with keys $k_i, \ldots, k_j$, and the dummy keys $d_{i-1}, \ldots, d_j$.

We associate a cost $c[i, j]$ with this binary search subtree. We minimize $c[i, j]$. By the principle of optimality, we observe that if an OBST, T has a subtree T with keys $k[i], \ldots, k[j]$ and dummy keys $d[i-1], \ldots, d[j]$, then T itself is optimal for the subproblem. For 'cost' minimization of the search, if by chance, there is another subtree T'' whose 'expected cost' is less than T, we discard T and consider T''.

Given an optimal subtree T with keys $k[i], \ldots, k[j]$, let the node with key $k[r]$, $(i \leq r \leq j)$ be the root of T. Naturally, the left subtree of the root of $k[r]$ will contain keys, $k[i], \ldots, k[r-1]$ and the dummy keys $d[i-1], \ldots, d[r-1]$, while the right subtree contains the keys $k[r+1], \ldots, k[j]$ and the dummy keys $d[r], \ldots, d[j]$. We examine all candidate root keys $k[r]$. We generate all possible optimal (left or right) binary search subtrees. We finally obtain OBST.

Let $$w[i, j] = \sum_{l=i}^{j} p_1 + \sum_{l=i-1}^{j} q_1 \qquad (5.7)$$

Hence, if k_r is the root node of the subtree with keys $k_i, \ldots, k_j$, then

$$c[i, j] = p_r + (c[i, r-1] + w[i, r-1] + c[r+1, j] + w[r+1, j]\} \qquad (5.8)$$

Equation (5.7) gives

$$w(i, j) = w(i, r-1) + p_r + w(r+1, j) \qquad (5.9)$$

In view of this, Eq. (5.8) can be rewritten as

$$c[i, j] = \begin{cases} q_{i-1} & \text{if } j = i-1 \\ \min\limits_{1 \le r \le j} \{c[i, r-1] + c[r+1, j] + w[i, j]\} & \text{if } j \ge i \end{cases} \qquad (5.10)$$

For initializing the optimal search, we define root $[i, j]$ to be r, the index of the root node $k[r]$. Also, we take

$$w[i, i-1] = q_{i-1}, \text{ for } 1 \le i \le n \text{ and}$$
$$w[i, j] = w[i, j-1] + p_j + q_j \text{ for } j \le i$$

The necessary algorithm for the OBST search is given below:

Algorithm 5.3 Minimum cost BST

```
Algorithm OBSTree(p, q, n)
// This algorithm carries out the optimal search in a given
binary
// search tree. a[1 : n] is a increasing sequence of identifiers
// p[1 : n] and q[0 : n] are probabilities that the search is for
the key
// k[i]. D[i] is a dummy key.
// This algorithm computes the cost c[i, j] of optimal binary
search
// tree tij for ai+1, ..., aj. This algorithm also returns the
// root, r[i, j] of tij. Weight of the edge (i, j) is w[i, j]
{
    for j := 0 to n-1 do
    {
        c[j, j] := 0.0;
        w[j, j] := q[j];
        r[j, j] := 0 ;
        w[j, j+1] := q[j] + q[j+1] +p[j+1] ;
        r[j, j+1] := j+1 ;
        c[j, j+1] := q[j] + q[j, j+1] +p[j+1] ;
```

```
    }
    w[n, n] := q[n] ;
    r[n, n] := 0 ;
    c[n, n]: = 0.0 ;
    for k: = 2 to n do // Finds optimal trees with k nodes
       for j: = 0 to n-k do
          {
             i: = j+k;
             w[j, i]: = w[j, i-1] + p[i] +q[i];
             m: = Find(c, r, j, i);
             c[j, i]: = w[j, i] + c[j, m-1] + c[m, i];
             r[j, i]: = m;
          }
    write (c[0, n], w[0, n], r[0, n] );
}
```

```
Algorithm Find(c, r, j, i)
{
least: = ∞;
for k: = r[j, i-1] to r[j+1n i] do
   if (c[j, k-1] + c[k, i]) < least then
      {
         least: = c[j, k-1] + c[k, i];
         l: = k;
      }
   return l;
}
```

Example 5.7

Trace the above algorithm for the BST with $p[1...4] = [0.05, 0.2, 0.1, 0.05]$, $q[0...4] = [0.1, 0.2, 0.05, 0.05]$ and with identifiers a[1...4] = [end, goto, print, stop].

We have

$$w[i, i] = q[i], \quad c[i, i] = 0, \quad \text{and root}[i, i] = 0 \qquad (1)$$

$$w[i, j] = p[j] + q[j] + w[i, j-1]$$

$$c[i, j] = \min\{c[i, k-1] + c[k, j]\} + w[i, j] \quad \text{for } 1 < k \le j \qquad (2)$$

$$\text{root}\ [i, j] = \min\{c[i, j]\} \qquad (3)$$

Iteration-1:

$$w[0, 1] = p[1] + q[1] + w[0, 0] = 0.05 + 0.1 + 0.2 = 0.35$$

$$c[0, 1] = \min\{c[0, 0] + c[1, 1]\} + w[1, 1]$$

(Here $0 < k \le 1$, so $k = 1$ only in (2))

$$\text{root}\ [0, 1] = 1 = \min\{0 + 0\} + 0.35 = 0.35$$

Iteration-2:

$$w[1, 2] = p[2] + q[2] + w[1, 1] = 0.2 + 0.2 + 0.1 + 0.5 = 1.0$$

$$c[1, 2] = \min \{c[1, 1] + c[2, 2]\} + w[1, 2]$$

(Here $1 < k \le 2$, so $k = 2$ only in (2))

$$= \min \{0 + 0\} + 0.5 = 0.5$$

$$\text{root}[1, 2] = 2$$

Iteration-3:

$$w[2, 3] = p[3] + q[3] + w[2, 2] = 0.1 + 0.05 + 0.2 = 0.35$$

$$c[2, 3] = \min \{c[2, 2] + c[3, 3]\} + w[2, 3]$$

(Here $2 < k \le 3$, so $k = 3$ only in (2))

$$= \min \{0 + 0\} + 0.35 = 0.35$$

$$\text{root } [2, 3] = 3$$

Iteration-4:

$$w[3, 4] = p[4] + q[4] + w[3, 3] = 0.05 + 0.05 + 0.05 = 0.15$$

$$c[3, 4] = \min \{c[3, 3] + c[4, 4]\} + w[3, 4]$$

(Here $3 < k \le 4$, so $k = 4$ only in (2))

$$= \min \{0 + 0\} + 0.15 = 0.15$$

$$\text{root } [3, 4] = 4$$

Iteration-5:

$$w[0, 2] = p[2] + q[2] + w[0, 1] = 0.2 + 0.2 + 0.35 = 0.75$$

$$c[0, 2] = \min \{c[0, 0] + c[1, 2], c[0, 1] + c[2, 2]\} + w[0, 2]$$

(Here $0 < k \le 2$. So $k = 1, 2$ in (2))

$$= \min \{0 + 0.5, 0.35 + 0\} + 0.75 = 0.35 + 0.75 = 1.1$$

$$\text{root } [0, 2] = 2$$

Iteration-6:

$$w[1, 3] = p[3] + q[3] + w[1, 2] = 0.1 + 0.05 + 0.5 = 0.65$$

$$c[1, 3] = \min \{c[1, 1] + c[2, 3], c[1, 2] + c[3, 3]\} + w[1, 3]$$

(Here $1 < k \le 3$. So $k = 2, 3$ in (2))

$$= \min \{0 + 0.35, 0.5 + 0\} + 0.65 = 0.35 + 0.65 = 1$$

$$\text{root } [1, 3] = 2$$

Iteration-7:

$$w[2, 4] = p[4] + q[4] + w[2, 3] = 0.05 + 0.05 + 0.35 = 0.45$$

$$c[2, 4] = \min \{c[2, 2] + c[3, 4], c[2, 3] + c[4, 4]\} + w[0, 3]$$

(Here $2 < k \le 4$. So $k = 3, 4$ in (2))

$$= \min \{0 + 0.15, 0.35 + 0\} + 0.45 = 0.6$$

$$\text{root } [2, 4] = 3$$

Iteration-8:

$$w[0, 3] = p[3] + q[3] + w[0, 2] = 0.1 + 0.05 + 0.75 = 0.9$$

$$c[0, 3] = \min \{c[0, 0] + c[1, 3], c[0, 1] + c[2, 3], c[0, 2] + c[3, 3]\} + w[0, 3]$$

(Here, $0 < k \leq 3$, so $k = 1, 2$ and 3)

$$= \min \{0 + 1, 0.35 + 0.35, 1.1 + 0\} + 0.9 = 1.6$$

root[0, 3] = 2

Iteration-9:

$$w[1, 4] = p[4] + q[4] + w[1, 3] = 0.05 + 0.05 + 0.65 = 0.75$$

$$c[1, 4] = \min \{c[1, 1] + c[2, 4], c[1, 2] + c[3, 5], c[1, 3] + c[4, 4]\} + w[1, 4]$$

(Here, $1 < k \leq 4$, so $k = 2, 3$ and 4)

$$= \min \{0 + 0.6, 0.5 + 0.15, 1 + 0\} + 0.75 = 1.35$$

root [1, 4] = 2

Iteration-10:

$$w[0, 4] = p[4] + q[4] + w[0, 3] = (0.05 + 0.05).9 = 1$$

$$c[0, 4] = \min \{c[0, 0] + c[1, 4], c[0, 1] + c[2, 4], c[0, 2] + c[3, 4], c[0, 3] + c[4, 4] + w[0, 4]$$

(Here, $0 < k \leq 4$, so $k = 1, 2, 3$ and 4)

$$= \min \{0 + 1.35, 0.35 + 0.6, 1.1 + 0.15, 1.6 + 0\} + 1 = 1.95$$

root [0, 4] = 2, $k = 2$.

From the above computations, we observe that $c[0, 4] = 1.95$ is the minimum cost. This is for the root [0, 4] = 2. This implies that the root node is $a[2]$ =goto.

We take the root node as $N_{0,4}$ for $k = 2$, $a[2]$ = goto. Children of node $N_{i,j}$ are $N_{i,\, k-1}$, $N_{k,j}$.

QUESTIONS

1. Explain the method of solving the knapsack problem using dynamic programming approach. Solve the knapsack problem with $n = 3$, $m = 20$, $(p_1, p_2, p_3) = (25, 24, 15)$ and $(w_1, w_2, w_3) = (18, 15, 10)$.
2. Write algorithm for the optimal solution of the knapsack problem, using dynamic programming technique.
3. Apply the all-pairs shortest algorithm to the weighted graph whose adjacency matrix is:

$$\begin{bmatrix} 0 & 4 & 11 \\ 6 & 0 & 2 \\ 3 & \infty & 0 \end{bmatrix}$$

4. Explain the reliability problem. Derive its mathematical formulation and write algorithm for solving it.
5. Obtain optimal solution to the reliability problem with $c = 105$, $c = 15$, $c = 20$, $c = 20$, $r = 0.9$, $r = 0.8$, $r = 0.8$, and $u = 2$, $u = 3$, $u = 3$).
6. What is the Travelling Salesperson Problem? Explain the method of solving it.
7. Solve the Travelling Salesperson problem with the associated adjacency matrix being

$$\begin{bmatrix} - & 10 & 15 & 20 \\ 5 & - & 9 & 10 \\ 6 & 13 & - & 12 \\ 8 & 8 & 9 & - \end{bmatrix}$$

8. Explain the flow-shop scheduling problem. Present its mathematical formulation and discuss the method of its solution.
9. Write algorithm for the "Optimal Binary Search Tree" and discuss its time complexity.
10. Solve the OBST problem with ordered elements and with $p_1 = 1/4$, $p_2 = 1/8$, $p_3 = p_4 = 1/16$ and $q_0 = 1/4$, $q_1 = 3/16$, $q_2 = q_3 = q_4 = 1/16$. Discuss the reliability design and travelling salesperson problem using dynamic programming approach.

CHAPTER 6

Search and Traversal Techniques

6.1 INTRODUCTION

Solution of many search problems can be obtained by associating a binary tree with each problem. We search for a node or a subset of nodes of this binary tree having some specific properties. As such we traverse the binary tree. By saying that we visit a node, we may mean doing manipulations like (1) printing the contents of the node or (2) deleting the node or (3) editing the contents of the node or (4) searching for a node or (5) inserting a node if it is not present in the given binary tree.

6.2 BINARY TREE TRAVERSAL ALGORITHMS

There are three types of binary tree traversal methods which are as follows:

1. Pre-order binary tree traversal,
2. In-order binary tree traversal and
3. Post-order binary tree traversal method.

These three binary tree traversal methods are illustrated below. We suppose that each node has the following three components:

lptr	data	rptr

1. Pre-order binary tree traversal:
   (a) Process the node,
   (b) Traverse the left sub-tree of the node in 'Pre-order' and
   (c) Traverse the right sub-tree of the node in 'Pre-order'.

2. In-order binary tree traversal:
   (a) Traverse the left sub-tree of the node in 'In-order',
   (b) Process the node and
   (c) Traverse the right sub-tree of the node in 'In-order'.
3. Post-order binary tree traversal:
   (a) Traverse the left sub-tree of the node in Post-order;
   (b) Traverse the right sub-tree of the node in Post-order, and
   (c) Process the node.

We now give the concerned algorithms for the above.

Algorithm 6.1 Recursive pre-order binary tree traversal

```
Algorithm RPreorder (bt).
// bt is a pointer giving the address of the root of a binary tree.
//Each node of bt is a record having fields: lptr, data and rptr.
// This algorithm traverses the binary tree recursively in Pre-Order.
{
   if bt ≠ 0 then
   {
   visit(bt);
      if (lptr(bt) ≠ 0) then
         RPreorder(lptr(bt));
      if (rptr(bt) ≠ 0) then
         RPreorder(rptr(bt));
   }
   return;
}
```

Algorithm 6.2 Recursive in-order binary tree traversal

```
Algorithm RInorder (bt)
// bt is a pointer giving the address of the root node of the tree.
// Each node of bt is a record having fields: lptr, data and rptr.
{
   if (bt ≠ 0) then
   {
```

```
            if lptr(bt ≠ 0) then
               RInorder(lptr(bt);
            visit(bt);
            if (rptr(bt ≠ 0 ) then
               RInorder(rptr(bt));
         }
      return;
   }
```

Algorithm 6.3 Recursive post-order binary tree traversal method

```
Algorithm RPostorder (bt)
//bt is a pointer giving the address of the root node of a binary
tree.
// Each node of bt is a record having fields: lptr, data and
rptr.
{
   if (bt ≠) then
      {
         if lptr(bt ≠ 0) then
         RPostorder(lptr(bt));
         if rptr(bt ≠ 0) then
         RPostorder(rptr(bt));
         visit (bt) ;
      }
}
```

It can be seen directly that the time complexity of any of the above binary tree traversal algorithms is $T(n) = \Theta(n)$ and the space complexity is $S(n) = O(n)$.

6.3 THE REACHABILITY PROBLEM

Given a graph $G = (V, E)$ it is required to check whether there exists a path from a given node i to a given node j. This problem is called the reachability problem. The simplest way of representing a graph in the memory of a computer is through adjacency matrix representation. A graph with n nodes is represented by a $n \times n$ matrix $A = (a_{ij})$, where

$$a_{ij} = 1, \text{ if there is an edge between the nodes } i \text{ and } j$$
$$= 0, \text{ otherwise}$$

Matrix A is called the adjacency matrix of the graph G.

Example 6.1

Consider the following directed graph.

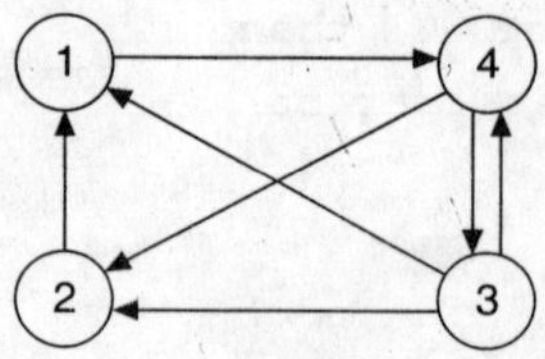

Figure 6.1 A directed graph.

The adjacency matrix is

$$A = \begin{bmatrix} 0 & 0 & 0 & 1 \\ 1 & 0 & 0 & 0 \\ 1 & 1 & 0 & 1 \\ 0 & 1 & 1 & 0 \end{bmatrix}$$

We now compute

$$B_r = A + A^2 + \dots + A^r \quad \text{for } r = 1, 2, \dots, \text{ where } A^k = A * A^{k-1}$$

$B_r(i, j)$ indicates the number of paths from i to j of length less than or equal to r.

If we want to know whether a path exists from node i to node j, we have to examine all elementary paths of length less than or equal to $n-1$. In case $i = j$ and the path is a cycle, we have to examine all elementary paths of length less than or equal to n. In the later case, we compute B_n, where

$$B_n = A + A^2 + \dots + A^n$$

For the above graph, we observe that

$$A^2 = \begin{bmatrix} 0 & 1 & 1 & 0 \\ 0 & 0 & 0 & 1 \\ 1 & 1 & 1 & 1 \\ 2 & 1 & 0 & 1 \end{bmatrix} \quad A^3 = \begin{bmatrix} 2 & 1 & 0 & 1 \\ 0 & 1 & 1 & 0 \\ 2 & 2 & 1 & 2 \\ 1 & 1 & 1 & 2 \end{bmatrix} \quad A^4 = \begin{bmatrix} 1 & 1 & 1 & 2 \\ 2 & 1 & 0 & 1 \\ 3 & 3 & 2 & 3 \\ 2 & 3 & 2 & 2 \end{bmatrix}$$

$$B_4 = A + A^2 + A^3 + A^4 = \begin{bmatrix} 3 & 3 & 2 & 4 \\ 3 & 2 & 1 & 2 \\ 7 & 7 & 4 & 7 \\ 5 & 6 & 4 & 5 \end{bmatrix}$$

We now define the path matrix of the graph as the matrix $P = \{p_{ij}\}$, where

$$P_{ij} = \begin{cases} 1, & \text{if there is a path from node } i \text{ to node } j \\ 0, & \text{otherwise} \end{cases}$$

One can obtain the path matrix P from B_n by setting $p_{ij} = 1$ if the element in the ith row, jth column is nonzero and $p_{ij} = 0$ otherwise. Thus a one (1) in the ith row, jth column means that there is a path from node i to node j. For the above graph, we observe that

$$P = \begin{bmatrix} 1 & 1 & 1 & 1 \\ 1 & 1 & 1 & 1 \\ 1 & 1 & 1 & 1 \\ 1 & 1 & 1 & 1 \end{bmatrix}$$

We consider the following two methods of solving the reachability problem.

6.3.1 Breadth First Search (BFS)

This is a search method in a graph. We begin our search with a node. We mark it as having been visited. We say that this node is explored only if all the nodes adjacent to this node are visited. In BFS, all nodes not visited adjacent to this node are visited next. We maintain a queue for keeping track of the adjacent nodes of a node visited. The top node is the next node to be visited. Following is an algorithm 6.4 for BFS technique.

Algorithm 6.4 Breadth first search

```
Algorithm BFS(v)
// G is a given graph.
// The BFS begins at node u.
// Value of the element status[i] is set to 1 if node i visited.
// All the array elements status[ ] are initially set to zero.
// Q is a queue of unexplored nodes
  {
     t := u ;
        status[u] := 1 ;
        repeat
           {
              for all nodes w adjacent to t do
                 {
                    if ( status[w] = 0 ) then
                       {
                          Insert w into Q ;
                          status[w] := 1 ;
                       }
                 }
                 if Q is empty then return ;
                 Delete t from Q ;
           }
                 until (false) ;
  }
```

A graph G is said to be a connected graph if there is at least one path between every pair of its vertices. Otherwise, the graph is said to be unconnected. When BFS method is applied to a connected graph G then all the vertices are visited, and we can claim that the graph is traversed. If G is not connected, then some k, $(1 \le k \le n)$ vertices are not visited. Interestingly, a repeated application of BFS, each time starting from an unvisited node, allows for the tree traversal.

This aspect is explained in the following algorithm 6.5.

Algorithm 6.5 Breadth first traversal of a graph

```
AlgorithmBFT(G)
// G is an un-connected graph with n vertices.
// If vertex j is visited then the value of the element status[j]
is set to 1.
// We set initially status[j] = 0 for 1 ≤ j ≤ n.
{
for j: = 1 to n do
status[j]: = 0;
for j: = 1 to n do
if (status[j] = 0) then
BFS(j);
}
```

6.3.2 Depth First Search (DFS)

In this search method when a node is visited, then all its adjacent nodes are stacked. Then the top element of the stack is processed and again all its adjacent nodes are pushed into that stack and process continues. Note that the first node adjacent to the first processed node is the last node to be processed. We now give an algorithm 6.6 for the DFS.

Algorithm 6.6 Depth first search

```
AlgorithmDFS(v)
// G = (V, E) is the given graph.
// G has n vertices. We assign 1 to status[i] if the node I is
visited.
// Initially all elements of status[] are set to zero.
{
   status[v]: = 1;
      for each vertex w adjacent to v do
         {
            if (status[w] = 0) then DFS(w);
         }
}
```

6.4 CONNECTED COMPONENTS AND SPANNING TREES

The BFS and the BFT algorithms are applied to test whether a graph G is connected or not and to obtain the connected components of G. Consider the set of all edges (u, w), where all vertices w are adjacent to u and are not visited. Refer to the BFS algorithm. It is established that this set of edges give the spanning tree of G, if G is connected. We obtain depth first search spanning trees similarly.

Consider the graph as shown in Figure 6.2.

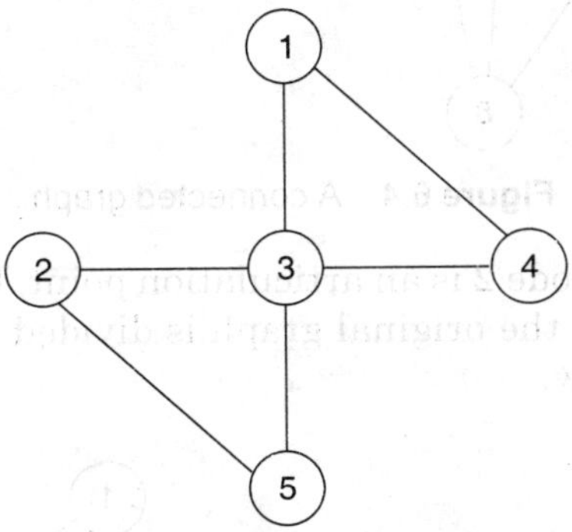

Figure 6.2 A graph.

The following are the DFS and BFS spanning trees of the graph of Figure 6.2.

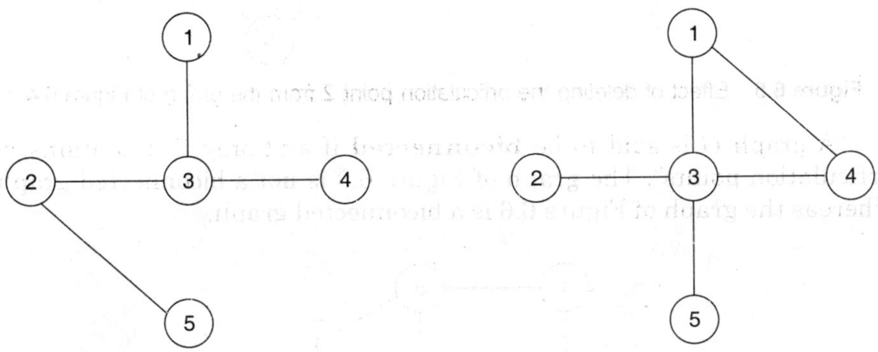

(a) DFS spanning tree (b) BFS spanning tree

Figure 6.3 DFS and BFS spanning trees of the graph of Figure 6.2.

6.5 BICONNECTED COMPONENTS

Let $G = (V, E)$ be an undirected graph. "A vertex v in a connected graph G is called an **articulation point** if and only if deletion of that vertex v together with all edges incident with V disconnects the graph into two or more nonempty components".

Consider the connected graph:

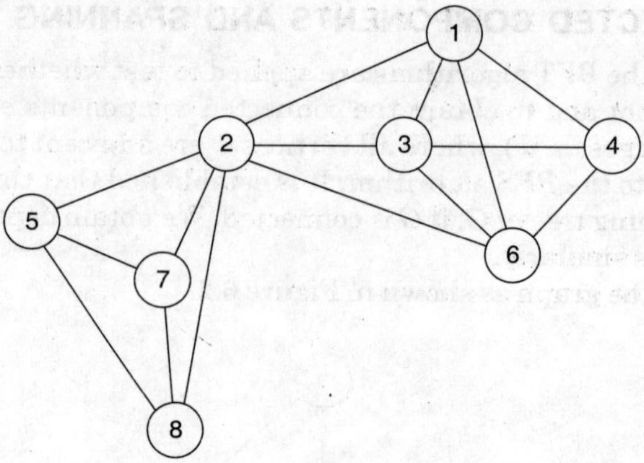

Figure 6.4 A connected graph.

We observe that node 2 is an articulation point. If this node is deleted, then the effect is that the original graph is divided into two independent graphs as shown below.

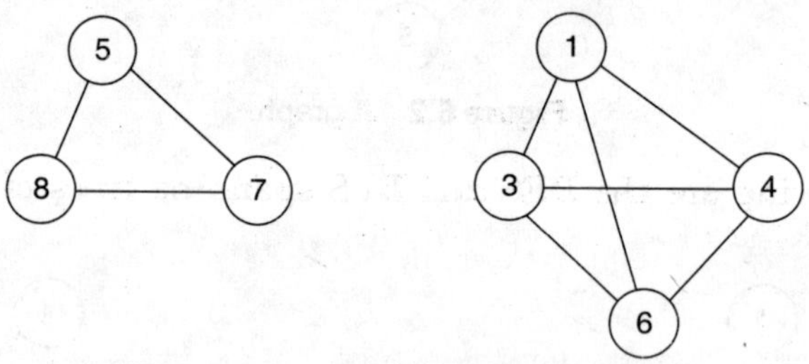

Figure 6.5 Effect of deleting the articulation point 2 from the graph of Figure 6.4.

"A graph G is said to be **biconnected** if and only if it contains no articulation points". The graph of Figure 6.3 is not a biconnected graph, whereas the graph of Figure 6.6 is a biconnected graph.

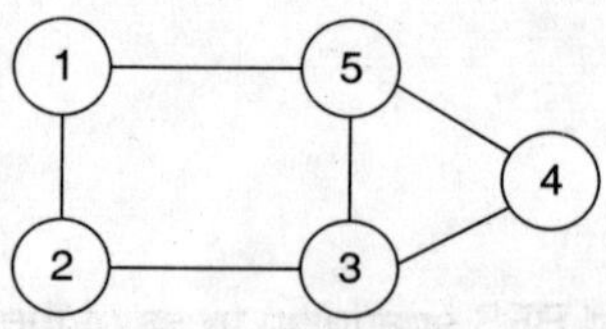

Figure 6.6 A biconnected graph.

Now the problem is to check whether a given graph is a biconnected one or not and to identify the articulation points if the graph is not a biconnected one. Before we attempt writing algorithms for these two problems, note the following results:

"Two biconnected components can have at most one vertex in common and this vertex is an articulation point".

"Every biconnected component of a connected graph contains at least two vertices, unless G has a single node".

The biconnected components obtained by edge addition of graph of Figure 6.3 are given in Figure 6.7.

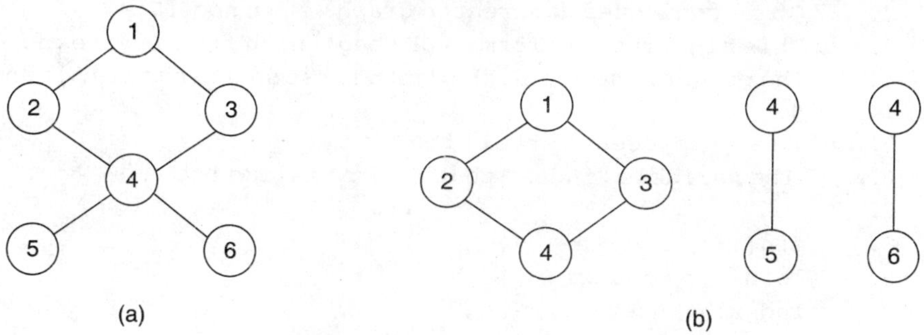

Figure 6.7 Graph (a) and its biconnected components.

Consider the following graph and its depth first spanning tree.

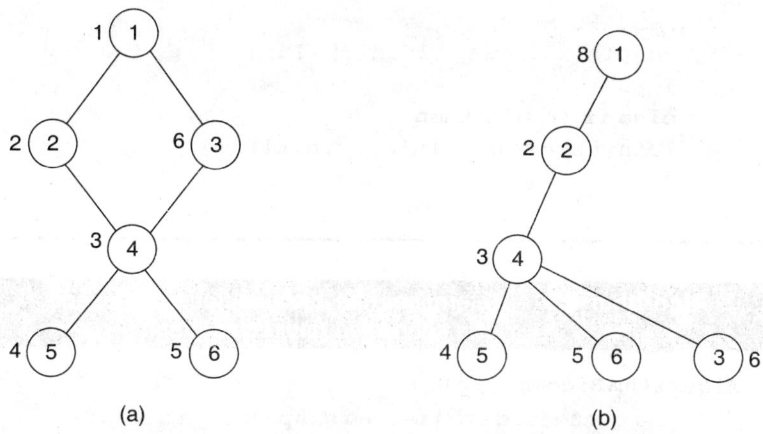

Figure 6.8 A graph and its depth first spanning tree.

For identifying the articulation points of a graph, the depth first spanning tree of G is used.

The side numbers tell the order in which a DFS visits these vertices. These numbers are called the depth first numbers (dfns). If i is a vertex in the graph, and k is a child of i

$$\text{ldfn}[i] = \min[\text{dfn}[i], \min\{\text{dfn}[k] \mid k \text{ being a child of } i, \min\{\text{dfn}[j] \mid (i, j) \text{ being a back edge}\}\}$$

Vertex i will be an articulation point if

1. i is not a root, and
2. $\text{ldfn}[k] \geq \text{dfn}[i]$.

In fact, among all the vertices that can be reached from vertex i, ldfn[i] defines the least dfns, followed by at most one back edge.

The following algorithm performs DFS on G and computes L(u):

Algorithm 6.7 Computes dfn and ldfnL

```
Algorithm ArtPoint(i, j)
// G is a connected undirected graph with n nodes.
// This algorithm performs a DFS beginning from a vertex i.
// The parent node of i in the depth first spanning tree, if any,
is j.
// k is an adjacent vertex of i.
// The variable "index" that is initialized to 1.
{
    dfn[i]: = index;
    ldfn[i]: = index;
    index: = index + 1;
    for each node k adjacent to u do
    {
        if (dfn([k] = 0 ) then
        {
            ArtPoint(k, i);
            ldfn[i]: = min( ldfn[i], ldfn[k] );
        }
        else if (k ≠ j ) then
        ldfn[i] :=min( ldfn[i]. dfn[k]);
    }
)
```

Algorithm 6.8 Determines the biconnected components of a connected undirected graph

```
Algorithm Bicomp(i, j)
// G is a connected undirected graph with n nodes.
// This algorithm computes biconnected components of G by
// performing a DFS beginning from a vertex i.
// Node j is the parent node of u in the depth first spanning tree.
// k is an adjacent vertex of i.
// Variable "index" is initialized to 1.
{
    dfn[i]: = index;
    ldfn[i]: = index;
    index: = index + 1;
        for each vertex k adjacent to i do
            {
            if ((j ≠ k) and (dfn[k] < dfn[i])) then
                Push (i, k) to a stack S;
```

```
            if (dfn[k] = 0) then
              {
                 if (ldfn[k] ≥ dfn[i]) then
                    {
                       write ("New bicomponent");
                       repeat
                          {
                             POP the top edge (x, y) from the stack S;
                             write ((x, y));
                             }until (((x, y) = (i, k)) or ((x, y) = (k, i)));
                          }
                          BiComp(k, i);
                          ldfn(i) := min(ldfn[i], ldfn[k]);
                    }
                    else if ( k ≠ j ) then
                    ldfn[i] := min(ldfn[i], dfn[k]);
              }
}
```

QUESTIONS

1. Explain and write algorithms for (a) Pre-order, (b) In-order, and (c) Post-order binary tree traversals.
2. Explain the BFS method with an example.
3. Write the BFS algorithm.
4. Explain the DFS method with an example.
5. Write the DFS algorithm.
6. Write about connected graphs and spanning trees. What are their advantages?
7. Define a biconnected graph. Write the algorithm for identifying the biconnected components of a graph.

CHAPTER 7

Backtracking Methods

7.1 INTRODUCTION

Backtracking is a very useful technique used in the design of algorithms. This technique generates all possible solutions that satisfy a given set of conditions. Any such solution is an n-tuple $(x_1, x_2, \ldots, x_n)$, where the values of x_i are members of some finite set S_i. We identify a "criterion function", also called "bounding function", $P(x_1, x_2, \ldots, x_n)$ and the problem is to find a vector $(x_1, x_2, \ldots, x_n)$ or a set of such vectors that maximizes (or minimizes or satisfies) the criterion function. From the different arrangements of the given elements, we select the required sequence. For example, let $a[1 : 5]$ be an unsorted array of 5 elements. The sorted sequence can be expressed by a vector say, (3, 1, 5, 2, 4) indicating that $a[3] \le a[1] \le a[5] \le a[2] \le a[4]$.

Consider $|S_i| = s_i$. Then there are $m = s_1, s_2 \ldots s_n$, solutions, each a n-tuple, that satisfy all this criteria of the function P. The simple and straightforward working rule is to list all possible solutions and then select the required solutions that satisfy the optimum criterion. But this method is tedious and time-consuming. The advantage of the backtracking method is that the answer can be obtained in less than m trials. In this method, we generate the solution vector by adding one component at a time. We use auxiliary criterion functions $P_i(x_1, x_2, \ldots, x_i)$. These auxiliary bounding functions are used to test what next element to be included in the partial solution that is generated by them. If it is found that the partial vector $(x_1, x_2, \ldots, x_i)$ does not lead to an optimal solution, then $s_{i+1}, \ldots s_n$ possible solutions need not be considered for inclusion into the set of feasible solutions.

The constraints that the solutions have to satisfy are classified into two categories: **explicit** and **implicit**.

Definition (Explicit Constraints): The constraints x_i can take values from S_i. These constraints depend upon the particular instance of the problem. Some examples of explicit constraints and their corresponding sets are given below.

Constraint	*Set*
$x_i = 0$ or 1	$S_i = \{0, 1\}$
$0 \le x_i \le 1$	$S_i = \{v \mid 0 \le v \le 1\}$

Definition (Implicit Constraints): The implicit constraints identify the tuples that are to be included in the solution space of i, satisfying the criterion function.

Before applying the backtracking technique to some problems, first we present an algorithm for generating all the possible permutations of a given set of n elements.

7.2 RECURSIVE PERMUTATION GENERATOR

Given a set of n or more elements, the following algorithm generates all possible permutations of elements of this set. Finding all such permutations will be useful in solving search problems. We consider this aspect later.

Algorithm 7.1 Recursive algorithm for permutation generator

```
Algorithm Permutations (a, i, n)
//a[] is an array of size n
// This algorithm generates all possible permutations of the
// n elements of the array a[].
{
   if (i = n) then write (a[1:n]); // Only one permutation.
   else
      for j = i to n do
   {
      temp: = a[i];
      a[i]: = a[j];
      a[j]: = temp;
         Permutations (a, i + 1, n);
         temp: = a[i];
         a[i]: = a[j];
         a[j]: = temp;
   }
}
```

Backtracking algorithms generate the problem solutions by searching the solution space for the given problem instance in some order. For this purpose **a tree organization** for the solution space is adopted. In this chapter, we consider the following problems and solve them using backtracking technique: (1) the 8-queen problem in a chess board, (2) the n-Queen problem, (3) the sum of subset problem, (4) the graph colouring problem, (5) the Hamiltonian cycles problem, and (6) the knapsack problem, etc.

7.3 THE 8-QUEEN PROBLEM

Take a chess-board. As we know that it contains 64 cells or squares (having two colours, viz. black and white alternatively) arranged in 8 rows and 8 columns. The problem is to place 8 queens in the chessboard squares so that no two of them attack each other, that is no two of them are on the same row, column, or diagonal. For future reference we put the numbers 1, 2, ..., 8 to the rows (from left to right) and columns (from top to bottom) of the chessboard as shown below. We also label the queens as 1, 2, ..., 8. There will be no loss of generality if we assume that queen i is placed on row i. The solution set to the 8-queen problem is then the set of all 8-tuples $(x_1, ..., x_8)$, where $1 \le x_i \le 8$ and x_i is the column on which queen i is placed.

1. The explicit constraints of the problem are $S_i = \{1, 2, 3, 4, 5, 6, 7, 8\}$, for $1 \le i \le 8$ and
2. The implicit constraints are that neither two queens are on the same column nor on the same diagonal.

The solution space consists of 8^8 elements, each of 8-tuples. The explicit constraints reduce the number of possible solutions to be the permutations of the 8-tuple (1, 2, 3, 4, 5, 6, 7, 8). This observation reduces the solution space from 8^8 tuples to 8!. One valid solution to the 8-queen problem is: (4, 6, 8, 2, 7, 1, 3, 5) and is shown in Figure 7.1.

	1	2	3	4	5	6	7	8
1					*Q*			
2		*Q*						
3						*Q*		
4	*Q*							
5							*Q*	
6				*Q*				
7								*Q*
8			*Q*					

Figure 7.1 Solution to the 8-queens problem.

7.4 THE *n*-QUEEN PROBLEM

This is a generalization of the 8-queen problem. Imagine a $n \times n$ chess-board. The problem is to place on this board all the n queens so that no two can attack each other. As discussed earlier, the solution space consists of as many as $n\,!$ permutations of the n-tuple $(1, 2, \ldots, n)$. We solve the n-queens problem using backtracking technique. Let $(x_1, x_2, \ldots, x_n)$ be a solution in which x_i is the column number of the ith row in which the ith queen is placed. Thus $x_6 = 4$ means that the place of the 6th queen is row 6 and column 4. Observe that no two x_i's are equal since no two queens can be placed in the same column. We finally test whether two queens are on the same diagonal.

Suppose the squares of the $n \times n$ chess-board are numbered as the indices of the two dimensional array $a[1 : n, 1 : n]$ $a[1, 1]$ is the top left hand square and $a[n, n]$ is the bottom right hand corner square), then we observe that every element on the same diagonal that runs from the upper left to the lower right has the same row-column value.

In Figure 7.1 (of 8-queen problem), consider the queen at $a[4, 3]$. The squares that are diagonal to this queen from the upper left to the bottom right are $a[2, 1]$, $a[3, 2]$, $a[5, 4]$, $a[6, 5]$, $a[7, 6]$, and $a[8, 6]$. Observe that the differences of the row and column indices of all these squares are same (= 1). Also, elements from upper right to bottom left corresponding to square $a[4, 3]$ are: $a[1, 6]$, $a[2, 5]$, $a[3, 4]$, $a[5, 2]$ and $a[6, 1]$. Observe that in all these squares the sum of row index and the column index is same (= 7). Mathematically speaking, two queens placed in squares (i, j) and (k, l) will be on the same diagonal only if

$$i - j = k - 1 \quad \text{or} \quad i + j = k + 1$$

These two conditions imply

$$j - 1 = i - k \text{ and } j - 1 = k - i$$

Thus, two queens will be the same diagonal if and only if $|j - 1| = |i - k|$.

We now give two algorithms necessary for solving this n-queens problem.

Algorithm 7.2 n-queens problem—Place function

```
Algorithm Place (i, j)
// Returns true if a queen can be placed in ith row and jth
// column. abs ( ) is the absolute function.
{
  for k = 1 to i - 1 do
    if ((x[k] = j ) or (abs(x[k] - j))
    then return false;
    return true;
```

Algorithm 7.3 All solutions to the n-queens problem

```
Algorithm nQueens(i, n)
// Prints all possible placements
// of n queens on an n x n chess-board so that they are
   nonattacking.
{
  for j = 1 to n do
  {
    if Place(i, j) then
    {
      if (i = n) then write (x[1:n]);
      else nQueens(i + 1, n);
    }
  }
}
```

In algorithm 7.2, Place (i, j) returns a boolean value that is true if the ith queen can be placed in column j. It tests both whether j is distinct from all previous values $x[1], \ldots, x[i-1]$ and whether there is no other queen on the same diagonal. Its computing time is $O(k-1)$.

We now use Place in algorithm 7.3 and obtain a solution to the n-queens problem. The algorithm is invoked by n-Queens $(1, n)$.

The following terminology regarding tree organizations of solution space is used.

Each node in the aforesaid tree defines a **problem state**. All paths from the root to other nodes define the **state space of the problem.**

Solution states are those problem states i for which the path from the root to i defines a tuple in the **solution space.**

Answer nodes are those solution states i such that the paths from the root to node i is a tuple that satisfies the implicit condition in addition and as such is a member of the set of solutions of the problem.

The above considered tree is also called as the **state space tree.**

A **live node** is one that is generated but all of whose children are not yet generated.

The **live node** whose children are currently being generated is called the **E-node**.

A **dead node** is a node that is generated for which no further expansion is possible (may be it is leaf node or all its children are generated). Consider the following illustrations.

Some special functions called the **bounding functions** are used to discard (kill), if necessary, without generating all their children.

(1)

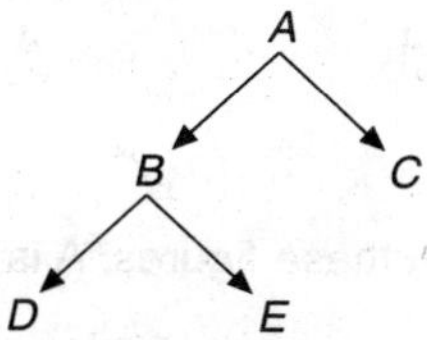

(a) ABD, ABE, AC are tuples in the solution space.

(2)

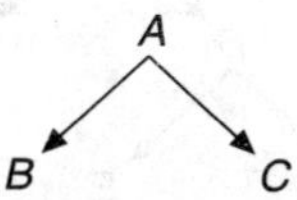

(b) A, B, C are problem states

(3)

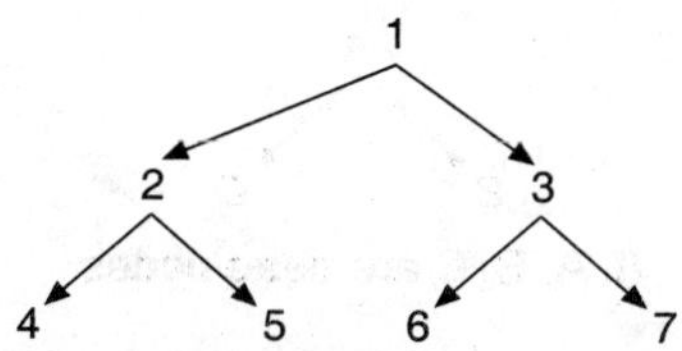

(c) (1, 2, 4), (1, 3, 6), (1, 3, 7) are solution states

(4)

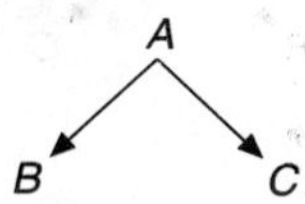

(d) Node A is not a live node. Nodes B and C are live nodes

(5)

A

(e) A is a live node

(6)

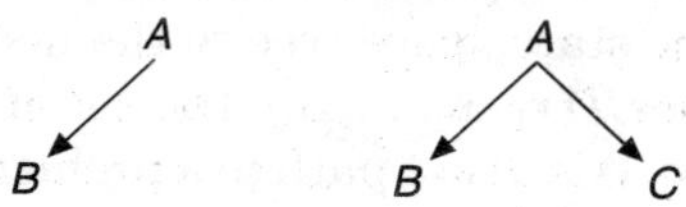

(f) and (g) In both these figures, A is not an E-node

(7)

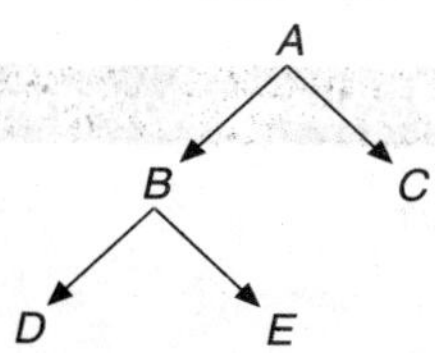

(h) D, E, C are live nodes

(8)

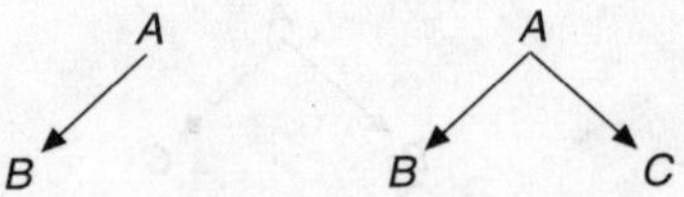

(i) and (j) In both these figures, A is not an E-node

(9)

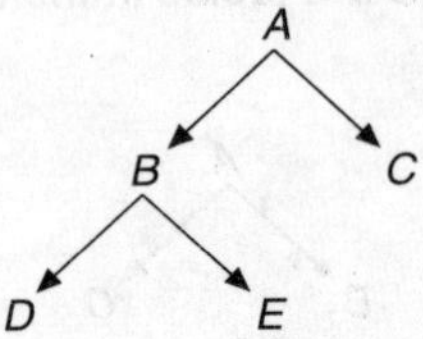

(k) E is not an E-node.

(10)

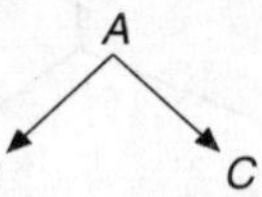

(l) A, B, C are dead nodes

(11)

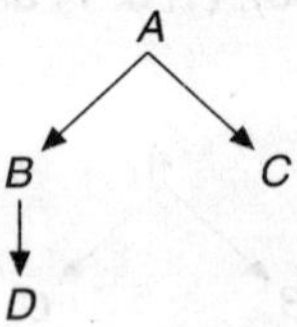

(m) A, D, C are dead nodes

Figure 7.2 Different types of nodes.

We now discuss in general the above formulation of backtracking problem of finding all answer nodes. The recursive algorithm, given below, works as follows. Suppose at any instance, $(x_1, x_2, \ldots, x_i)$ be a path from the root node to a node in the state space tree to the node x_i. We now consider the node x_{i+1}. We consider $T(x_1, x_2, \ldots, x_i)$, the set of all possible values for x_{i+1} such that $(x_1, x_2, \ldots, x_{i+1})$ is also a path to a problem state. x_{i+1} is included in the solution vector only if B_{i+1} is true and is generated by T only. We observe that $T(x_1, x_2, \ldots, x_n) = j$.

We now give a recursive algorithm for the backtracking technique.

Algorithm 7.4 Recursive algorithm for backtraking

```
Algorithm RBacktrack(i)
// Suppose (x[1], x[2], ..., x[i-1]) is a path from the root in
a state
// space tree. This algorithms determines the value of x[i],
so that (x[1],
```

```
// (x[1],x[2],..., x[i-1], x[i]) is an extended path.
{
   for (each x[i] εBT( x[1], x[2],..., x[i-1]) do
      {
         if (BFi(x[1], x[2],..., x[i) ≠ 0) then
            {
            if (x[1], x[2],..., x[i]) is a path to an answer node
               then write (x[1:n]);
                  if(i < n) then RBacktrack(i+1);
            }
      }
}
```

The above algorithm generates all solutions to be printed. Solution tuples of various sizes may be printed. If only one solution on the first occurrence of it is to be printed then a flag may be used. An iterative algorithm for backtracking is given below.

Algorithm 7.5 An iterative algorithm for backtracking

```
Algorithm ItBacktrack(n)
//This is an iterative algorithm that illustrates
   backtracking technique
//All possible solutions x[1 : n] are generated.
{
i: = 1;
while (i ≠ 0) do
{
if (there remains an untried x[i] εBT(x[1], x[2],..., x[i-1]not
yet tested existed) and
    BFi(x[1], x[2],..., x[i]) is true) then
  {
     if (x[1], x[2],..., x[i]) is a path to an answer node) then
        write (x[1 : i])
    i: = i+1;
  }
     else
        i: = i-1; // Backtrack
  }
}
```

Example 7.1

(4-queens problem): Consider the 4-queen problem and apply the above written algorithm of backtracking technique to solve it. The selection criteria for the bounding function sees that if (x_1, x_2, ..., x_i) is the path to the current E-node x_i, then among all children nodes x_{i+1} of x_i, we select that one which makes $(x_1, \ldots, x_{i+1})$ also a path (that is a feasible solution set). A partial state space tree (in fact the left subtree of the root) of this problem is shown in Figure 7.3.

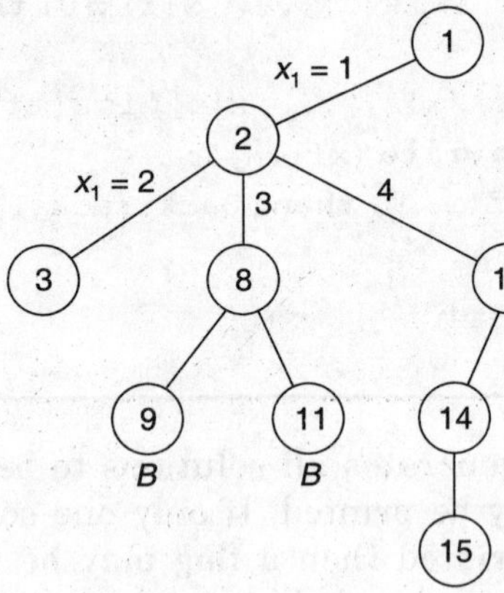

Figure 7.3 A partial tree (Left tree of the node 1) generated by backtracking for the 4-queen problem.

We start this search systematically, with the root node as the only live node. Now this root node becomes the E-node and the path is empty. We generate one child at a time. To avoid confusion, we generate children in ascending order of the parent-child labels. Node 2 is generated and the path is now (1). This suggests that queen 1 is to be placed in column 1 (Figure 7.4(a)). Node 2 becomes the E-node. Then node 3 is generated and immediately killed. Node 8 is generated next and the path is (1, 3). This suggests that queen 2 is to be placed in column 2 (Figure 7.4(b)). Now node 8 becomes the E-node. But it gets killed as none of its children can not lead to an answer node. We backtrack to node 2 and generate another child node 13. The path is (1, 4). We place queen 2 in column 4. This method is continued till a final solution is obtained. The tables in Figure 7.4 illustrate the different steps involved in the implementation of Algorithm 7.4.

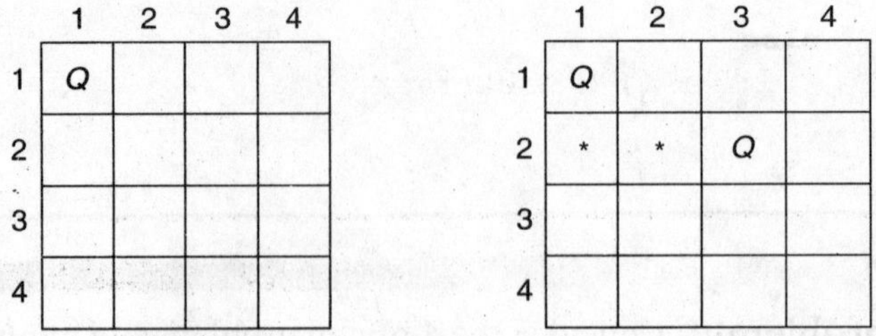

(a) (b)

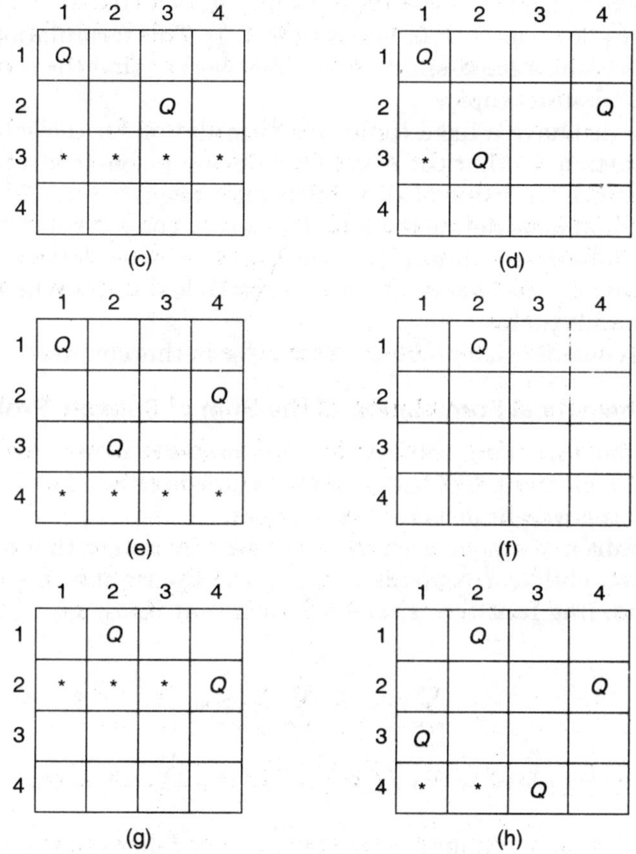

Figure 7.4 Solution to the 4-Queen problem.

7.5 SUM OF SUBSETS PROBLEM

We are given a set of n positive numbers w_i, $1 \le i \le n$, and a positive number m. The sum of subsets problem is to find all possible subsets of the given set of w_i's, whose sum is m. For example, if $n = 4$, $(w_1, w_2, w_3, w_4) = (12, 14, 26, 8)$, and $m = 34$, then the required subsets are {12, 14, 8} and {26, 8}. Sometimes the above two solutions are better shown by the indices of w_i, as the vectors (1, 2, 4) and (3, 4). Note that the sizes of these two solutions are different. In general, all solutions are k-tuples $(x_1, x_2, \ldots, x_k)$, for $1 \le k \le n$. Different solutions may have different sizes. The explicit constraints require $x_i \varepsilon \{j \mid j \text{ an integer and } 1 \le j \le n\}$. The implicit constraints are that no two solutions are identical but the sum of the corresponding w_i's is m. To avoid confusion between the solution vectors (1, 2, 4) and (1, 4, 2) and to treat them as identical, we lay down another implicit constraint as $x_i < x_{i+1}$.

No confusion arises if each solution subset is shown as a n-tuple: $(x_1, x_2, \ldots, x_n)$ such that each $x_i = 0$ or 1. $x_i = 0$ if w_i is not selected and $x_i = 1$

if w_i is selected. Then the size of any solution is n (fixed). The above two solutions are shown as (1, 1, 0, 1) and (0, 0, 1, 1). This formulation expresses all solutions using a fixed-sized tuples. We observe that the solution space consists of 2^n distinct tuples.

We can also have a fixed tuple size formulation for the solution space tree organization. In that the edges from level i nodes to level $i + 1$ nodes are labelled with the value of x_i, which is either zero or one. All paths from the root to a leaf node define the solution space. The left subtree of the root defines all subsets containing w_1, the right subtree defines all subsets not containing w_1, and so on. There are $2^4 = 16$ leaf nodes which represent different possible tuples.

We will consider this problem again later in this chapter.

7.5.1 Mathematical Formulation of the Sum of Subsets Problem

Consider a backtracking solution for this problem using fixed tuple size strategy. The element x_i of the solution vector is either 1 or 0, depending on whether the weight w_i is included or not.

The children of a node are now generated by noting that for a node at level i the left child corresponds to $x_i = 1$ and the right to $x_i = 0$.

The bounding function is selected such that $B_k(x_1, x_2, \ldots, x_k)$ = true if and only if

$$\sum_{i=1}^{k} w_i x_i + \sum_{i=k+1}^{n} w_i \geq m$$

$(x_1, x_2, \ldots, x_k)$ can lead to an answer node if and only if this condition is satisfied.

The bounding functions B_ks must satisfy the following two inequalities.

$B_k(x_1, x_2, \ldots, x_k)$ = true if and only if

$$\sum_{i=1}^{k} w_i x_i + \sum_{i=k+1}^{n} w_i \geq m \text{ and}$$

$$\sum_{i=1}^{k} w_i x_i + w_{k+1} \leq m$$

Now we present a recursive algorithm for the backtracking process for the sum of subsets problem.

Algorithm 7.6 Sum of subsets-recursive backtracking

```
Algorithm SumSubsets (sum, j, rw)
// This is a recursive algorithm that finds all solutions of
// the sum of subsets problem.
// The partial solution vector (x1, x2, …, xk-1) is already
selected.
```

```
// sum = Σ_{j=1}^{k-1} w[j] * x[j] and Σ_{j=k}^{n} w[j] denote respectively the
// cumulative sum up to (j-1) and the remaining weight left .
// The w[j]'s are in nondecreasing order.
// For all practical computations, we assume that w[1] ≤ m,
// and Σ_{i=1}^{n} w[i] ≥ m.
{
    // Generate left child.
        x[j]: = 1;
if (sum + w[j] = m) then write (x[1: j]);
// a feasible subset found.
else if ( sum + w[j] + w[j+1] ≤ m )
        then SumSubsets(sum + w[j], j+1, rw - w[j] );
// Generate right child.
if (( sum + rw - w[j] ≥ m) and (sum + w[j+1] ≤ m)) then
        {
           x[j]: = 0;
           SumSubsets(sum, j+1, rw - w[j]);
        }
}
```

Exercise

7.1 Solve the sum of subsets problem with $n = 6$, $w[1:6] = (5, 10, 12, 13, 15, 18)$, $m = 30$.

7.6 GRAPH COLOURING PROBLEM

This is a classical combinatorial problem. A graph G with n nodes and a positive integer m are given. The graph colouring problem is: using m colours only, to colour all the nodes of G in such a way that no two adjacent nodes have the same colour. This problem is called the m-colouring decision problem. The integer m is called the **chromatic number** of the graph.

We have the famous 4-colour planar graph problem. Given the map of the world in a plane sheet, the problem is to see whether all countries be coloured in such a way that no two adjacent countries have the same colour by using the given four colours only.

Each country is treated as a node of a graph and if two countries are adjacent to each other then there exists an edge between the two corresponding nodes.

For many years it was known that five colours were sufficient to colour any map, but no map that required more than four colours had ever been found. After several hundred years this problem was solved.

We represent a graph by its adjacency Boolean matrix $G[1 : n, 1 : n]$. The colours are represented by the numbers 1, 2, ..., m. The graph in Figure 7.5 can be coloured by 4 colours 1, 2, 3 and 4. $x[i]$ contains the colour of node i.

Solutions are n-tuples $(x_1, x_2, ..., x_n)$ where $1 \le x_i \le m$ for $1 \le i \le n$. x_i is the colour of the node i. The recursive backtracking algorithm is given below. The underlying space state tree used is a tree of degree m and height $n + 1$. Each node at level i has m children, corresponding to the m possible assignments to x_i, $1 \le i \le n$. Nodes at level $n + 1$ are the leaf nodes.

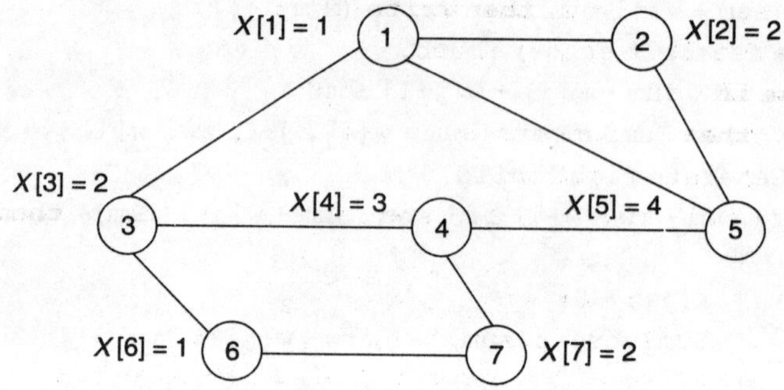

Figure 7.5 Graph colouring problem.

Solutions are n-tuples $(x_1, x_2, ..., x_n)$ where $1 \le x_i \le m$ for $1 \le i \le n$. x_i is the colour of the node i. The recursive backtracking algorithm is given below. The underlying space state tree used is a tree of degree m and height $n + 1$. Each node at level i has m children, corresponding to the m possible assignments to x_i, $1 \le i \le n$. Nodes at level $n + 1$ are the leaf nodes.

Algorithm 7.7 m-colouring problem

```
Algorithm mColouring(nv)
// This is a recursive algorithm for the m-colouring problem
// that uses backtracking technique.
// Graph G is maintained as a boolean adjacency matrix.
// No two adjacent vertices are assigned the same colour.
// nv is the index of the next vertex to colour
{
repeat
   {
      NextValue(nv);
      if (x[nv] = 0) then return; // Not possible to colour the vertex.
      if (nv = n) then // All nodes are coloured
```

```
        write (x[1 : n]);
        else m colouring(nv + 1);
    }
        until (false);
}
```

```
Algorithm NextValue(n-v)
// The vertices nodes 1, 2, . . ., nv-1 are coloured satisfying
the condition.
// That is x[1], x[2], . . ., x[nv-1] are assigned integer values
// from {1, 2, . . ., m}. X[k] is assigned a next highest numbered
colour.
// If we cannot find such a colour we set x[nv] to 0.
{
   repeat
      {
         x[nv]: = (x[nv]+1) mod (m+1);
         // Colouring next vertex.
         if (x[nv] = 0) then return;
         // No choice for a colour.
         for k = 1 to n do
         {
      // check if the colour is distinct from adjacent colours.
            if ((G[nv, k] ≤ 0) and (x[nv] = x[k]))
            if (nv,k) is an edge and if adjacent vertices have the
            same colour
               then break;
      }
         if (k = n + 1) then return;
            // New colour found.
      }        until (false);
            // Try another colour.
}
```

Example 7.2

Trace the above algorithm *m*-colouring for the graph and obtain its chromatic number.

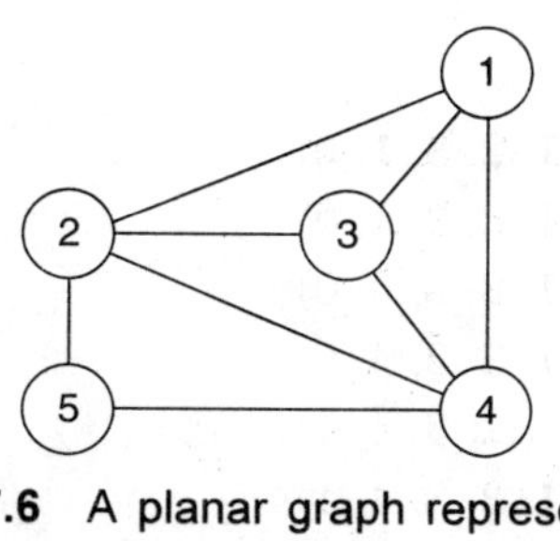

Figure 7.6 A planar graph representation.

Here $n = 5$, $m = n - 1 = 4$

Initially $x[\,]$ is set to zero. $x[\,] = (0,\ 0,\ 0,\ 0,\ 0)$

<u>For k = 1:</u> mColouring(1)

Call NextValue(1)

$$x[1] = x[2] \bmod (m + 1) \text{ gives } x[1] = 1.$$

Therefore, $x[\,]$ is: (1, 0, 0, 0, 0)

For $j = 1$ to 5 do

$$j = 1$$

If $G[k, j] \neq 0$ and $x[k] = x[j]$ then break gives, if $G[1, 1] \neq 0$ and $x[1] = x[1]$ gives (No and Yes), that is No.

Similarly, we observe that

For $j = 2 \rightarrow$ No

For $j = 3 \rightarrow$ No

For $j = 4 \rightarrow$ No

For $j = 5 \rightarrow$ No

<u>For k = 2:</u>

Call NextValue(2)

$$x[2] = (x[2] + 1) \bmod 5 \text{ gives } x[2] = 1$$

Vector $x[\,]$ becomes $x[\,] = 1,\ 1,\ 0,\ 0,\ 0$

Now the loop for $j = 1$ to 5 is applied and we see that Procedure NextValue is executed again.

$$x[2] = (x[2] + 1) \bmod 5 \text{ gives } x[2] = 2.$$

Vector $x[\,]$ is (1, 2, 0, 0, 0, 0)

Value of k is incremented by 1 in the mColouring function call.

<u>For k = 3:</u>

We obtain the vectors

(1, 2, 1, 0, 0)
(1, 2, 2, 0, 0)
(1, 2, 3, 0, 0)

<u>For k = 4:</u>

We obtain the vectors

(1, 2, 3, 1, 0)
(1, 2, 3, 2, 0)
(1, 2, 3, 3, 0)
(1, 2, 3, 4, 0)
(1, 2, 3, 4, 1)
(1, 2, 3, 4, 1)

7.7 HAMILTONIAN CYCLES

$G = (V, E)$ is a connected graph with n vertices (directed or undirected). A Hamiltonian cycle is a closed path that visits every vertex in V of G once and returns to its starting position. Suppose that the Hamiltonian cycle begins at vertex v_i of V and the vertices are labeled as $v_1, v_2, \ldots, v_{n+1} = v_1$, then the edges (v_i, v_{i+1}) are in E for $i = 1 \le i \le n$ and the vertices v_i are distinct except for v_1 which is equal to v_{n+1}. Consider the following graph.

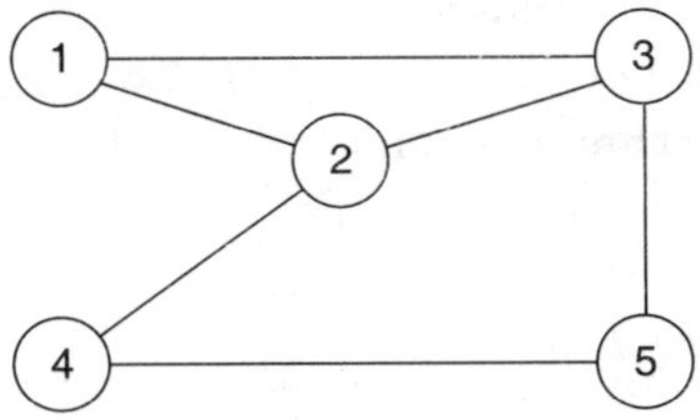

Figure 7.7 A graph containing Hamiltonian cycle.

The graph in the above figure contains the Hamiltonian cycle 1, 2, 8, 7, 6, 5, 4, 3, 1.

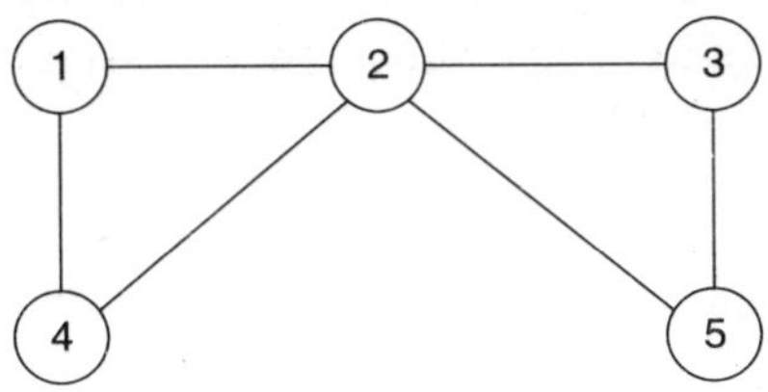

Figure 7.8 A graph containing no Hamiltonian cycles.

We now apply backtracking technique to find all the Hamiltonian cycles in a graph. The backtracking solution vector $(x_1, x_2, \ldots, x_n)$ is defined so that x_i contains the label of the node visited at ith visit. Now the problem is computing x_k if $x_1, x_2, \ldots, x_{i-1}$ have already been chosen. If $i = 1$, x_1 can be any of the n vertices. Without loss of generality, we set $x_1 = 1$. If $1 < i < n$, then x_i can be any vertex v that is distinct from $x_1, x_2, \ldots, x_{i-1}$ and v is connected by an edge to x_{i-1}. The vertex x_n must be connected to both x_{n-1} and x_1.

We first present the algorithm NextValue (i) which determines a possible next vertex for the proposed cycle.

We then use this algorithm in backtracking algorithm Hamiltonian(i), to find all Hamiltonian cycles.

Algorithm 7.8 All Hamiltonian cycles

```
AlgorithmNextValue(i)
// Identifies Hamiltonian cycles in a given graph G.
{
    repeat
```

```
    {
    x[i] := (x[i] + 1) mod (n + 1);
        //Next highest colour.
    if (x[i] = 0) then return;
    if (G[x[i-1], x[i]) ≠ 0) then              //Graph not null
        for k: = 1 to i-1 do
            if (x[k] = x[i]) then break;
            if (k = i) then
                if ((k < n) or ((k=n) and G[x[n], x[1]] ≠ 0))
                    then return ;
    }
  }
  until (false);
}
```

```
Algorithm Hamiltonian(i)
// This algorithm uses the recursive formulation of
//backtracking  to find all the Hamiltonian cycles of a graph.
// The graph is stored as an adjacency matrix G[1:n,1:n]. All
 cycles begin at node 1.
{
repeat
   {
      // values for x[i]
         NextValue(i);
      //is assigned a feasible value.
         if (x[i] = 0) then return ;
         if (i=n) then write(x[1: n]);
         else Hamiltonian(i+1);
   }
until(false);}
```

7.8 THE 0/1 KNAPSACK PROBLEM

We considered the 0/1 knapsack problem earlier. For effectiveness, we state it again now. We are given n weights w_i (> 0), weight w_i associated with a profit of p_i, and a positive number m, i.e. the knapsack capacity. The problem is to select a subset of the weights such that

$\Sigma w_i \, x_i \le m$ and

$\Sigma p_i \, x_i$ is maximized, i taking values 1, 2, ..., n.

x_i is 0 or 1

for all i, $1 \le i \le n$.

We recall our earlier discussion of this problem using the fixed tuple size formulation or the variable type tuple formulation. In either case, in applying backtracking technique to solve this problem, we define and use bounding functions to eliminate unwanted nodes. An upper bound on the value of the optimal solution is considered. Consider the fixed tuple formulation. Suppose the values of x_i for $1 \leq i \leq k$, have already been determined at node A. Then an upper bound for A can be obtained by noting that $x_i = 0$ or 1 to $0 \leq x_i \leq 1$, for $k + 1 \leq i \leq n$. An algorithm for the bounding function is given below.

Algorithm 7.9 Upper bound

```
Algorithm Bound(tcp, tcw, k)
// tcp is the total current profit, tcw is the current total
weight.
// k is the index of the last removed item.
{
   tb: = tcp;
   tc: = tcw;
   for j = k+1 to n do
      {
         tc: = tc + w[j];
         if (tc < m ) then tb: = tb + p[j]
            else return (tb + (1 - (tc - m)/w[j]) *p[j]);
      }
         return tb
}
```

Function Upper Bound (tcp, tcw, k) determines an upper bound on the optimal solution obtainable at level $k + 1$ of the state space tree. The object weights and profits are $w[j]$ and $p[j]$. It is assumed that $p[j]/w[j] \geq p[j + 1]/w[j + 1]$, for $1 \leq j \leq n$.

From the way the Bound is defined, we observe that the bound for a feasible left child of a node A is the same as that for A. Therefore, the bounding function need not be used whenever the backtracking algorithm makes a move to the left child of a node. In the following algorithm initially *fp* is set to –1 and the algorithm is invoked initially with a call to Bknap (1, 0, 0).

Algorithm 7.10 Backtracking solution of the 0/1 Knapsack problem

```
Algorithm BTknap(tcp, tcw, k)
// Applies backtracking technique to solve 0/1 Knapsack
   problem.
// tfp is the final total maximal profit.
// x[k] = 0 if w[k] is not in the knapsack, Otherwise x[k] = 1.
{
```

```
    //Generate left child of node k.
    if (tcw + w[k] ≤ m ) then
    {
      y[k]: = 1;
      if (k < n) then Bknap(tcp + p[k], tcw + w[k], k+1);
      if ((tcp + p[k] > tfp) and (k = n)) then
      {
        tfp: = tcp + p[k];
        tfw: = tcw + w[k];
        for i: = 1 to k do
          x[i]: = y[i];
      }
  }
    //Generate right child of node k.
      if (Bound(tcp, tcw, k) ≤ tfp) then
        {
          y[k]: = 0;
          if (k < n) then Bknap(k+1, cp, cw);
            {
              tfp: = tcp;
              tfw: = tcw;
              for i: = 1 to k do
              x[i]: = y[i];

        }
  }
```

All the problems discussed in this chapter and many such problems come under the category of "combinatorial search and optimization". In such methods an object from n given ones is selected to get an optimum output. This process is continued for the next object and so on, till the overall optimum is obtained. During this search process, if selection of a particular object does not improve the solution, we backtrack and select the another object. Consider the 8-queen problem. Note that the jth ($1 \le j \le 8$) queen is placed in the jth column of the chessboard. Suppose it is placed in the kth-row (of jth column) for $1 \le j \le 8$. If this selection violates the basic condition that no two queens attack each other, we then backtrack and change its position to another row in the same kth-column. Note that the kth-queen can be placed in any one of the 8 places. At each such placement a check for the optimality must also be carried out. This process is similar to searching a search tree for an optimum path. Consider the following set-up of a search tree.

If selection of the object O_{11} violates any or more of the given conditions, including the optimality, we backtrack to the previous object selected (O_j) and select the object O_{12}. This process of selection and testing and

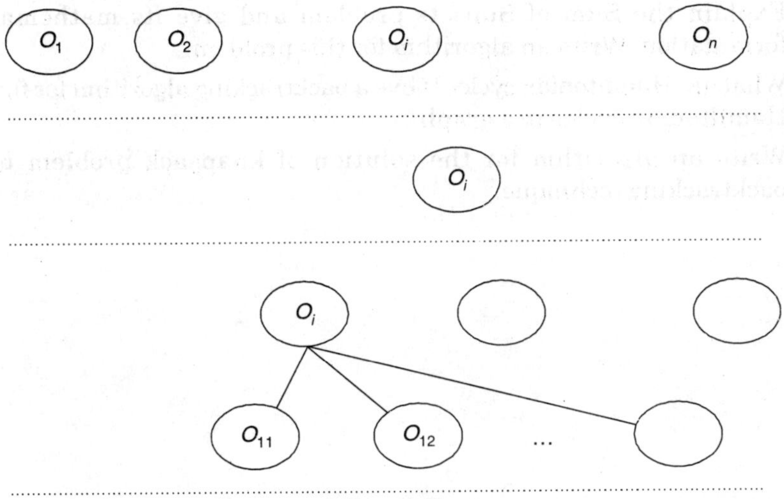

A Search Tres

backtracking if necessary is continued even to the level of selecting the first object. Search for the right object at the right stage is to be made. The following procedure for the combinatorial search for optimization of an object is given below and this is repeated for all objects.

```
1. Check if 'object' has all slots filled
2. If so, check controls and evaluate the objective function.
3. Check whether all constraints are satisfied
4. If so, return the 'object'.
   else
        return ∞
5. If object has unfilled slots  /* Do branching */
   Consider all possible ways of filling the next slot and
   let the object filled be 'i'.
   Let Vi be the modified objective function
     } /* end of loop */
     return Vi.
```

If you look at the above procedure, you identify this search algorithm with that of DFS.

QUESTIONS

1. Discuss the Backtracking procedure. Explain the explicit and implicit constraints of a backtracking problem.
2. Compare the brute force method with the backtracking method.
3. Write the explicit and implicit constraints of the 8-Queen problem and discuss its solution.
4. What is the graph colouring problem? Discuss in detail the m-colouring graph problem. Draw the state space tree for the m-colouring problem.

5. Explain the Sum of Subsets problem and give its mathematical formulation. Write an algorithm for this problem.
6. What are Hamiltonian cycles? Give a backtracking algorithm for finding Hamiltonian cycles in a graph.
7. Write an algorithm for the solution of knapsack problem using backtracking technique.

CHAPTER 8

Branch-and-Bound Methods

8.1 INTRODUCTION

Branch-and-Bound technique uses a slight variation of backtracking technique. Branch-and-Bound involves state space search methods wherein all children of an E-node are first generated before any other live node can become an E-node. When the search from a node stops, we get an **answer node.** Commonly used state search strategies are FIFO, LIFO and LC (Least Cost).

In the terminology of branch-and-bound, a BFS search state space is called the FIFO. A D search state space is called a LIFO. The corresponding algorithms are called "FIFOBB" (First in First out Branch and Bound) and "LIFOBB" (Last in First out Branch and Bound), and "LCBB" (Least Cost Branch and Bound) search methods.

8.2 BRANCH-AND-BOUND SEARCH METHODS

In this section, we consider different branch-and-bound search methods.

8.2.1 FIFO Branch and Bound (FIFOBB) Search

Consider the following graph as shown in Figure 8.1.

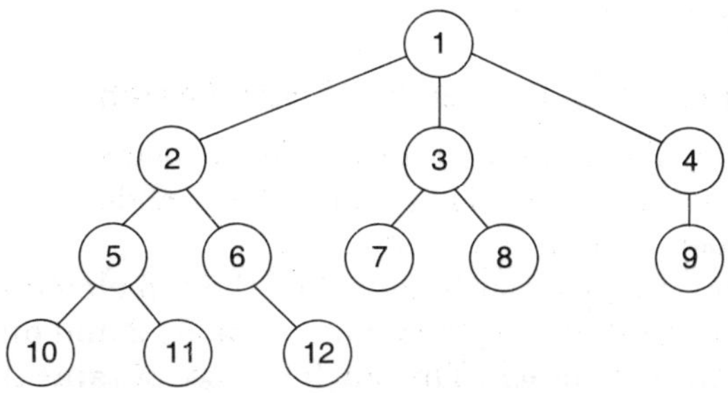

Figure 8.1 FIFO Branch-and-Bound.

Assume that node (12) is an answer node and all others are killed. Initially, node (1), the root node, is the E-node and the queue is empty.

We now generate the children nodes 2, 3 and 4 of node 1. We place all these children in a queue.

Queue

2	3	4	...

We delete the first element, node (2) of this queue and generate all its children, namely nodes 5 and 6 and put them in the queue.

3	4	5	6	...

Now delete the top element, node (3) of this queue and generate all its children nodes (7) and (8). As these nodes (7) and (8) are already killed they are not entered into the queue. As such now the queue is

4	5	6	...

Node (4) becomes E-node. It has only one child, namely node (9). Generate it. Delete the E-node. As node (9) is killed, it will not enter into the queue. The queue now is:

5	6	...

Now node (5) is the E-node. Generate its children, namely nodes (10) and (11). Delete the E-node (5) from the queue. As the children nodes are killed nodes, they will not enter into the queue. The queue now is

6	...

Node (6) becomes the E-node. Generate its only child, node (12). Delete the top element, node (6) from the queue and enter node (12) into the queue.

12	...

As there are no children of node (12), the search stops. So the optimal path in First in First out Branch and Bound (FIFOBB) is:

$$1 \rightarrow 2 \rightarrow 6 \rightarrow 12$$

8.2.2 LIFO Branch and Bound (LIFOBB) Search

Consider again the above directed graph. We consider a stack S which is initially empty. The live node (1) becomes the E-node. Generate its nodes (2), (3), (4) and push them into the stack S.

The stack now is 2, 3, 4 (with node 2 as the top element).

Pop the top element of the stack, namely node 2 and push its children nodes (5) and (6) into the stack. The stack S now contains the nodes 5, 6, 3 and 4 with node 5 on the top. We continue this process. We see that the elements (nodes) that nodes (5), (6) will be popped and at the last stage the stack contains the elements (nodes): 12, 3 and 4. As node (12) is the answer node, the search stops. So, the LILOBB search is: 1, 2, 5, 6 and 12.

8.2.3 LC Branch and Bound (LCBB) Search

Among different competing nodes, suppose there is a better chance for a particular node to become an *E*-node. If that particular node is selected, the answer node can be obtained quickly. This policy cannot be implemented either in the FIFOBB or in the LIFOBB searches, because the selection rule for the next *E*-node is sequential and all children of the present *E*-node are to be generated before any other live node can compete as a *E*-node. Lot of search time may be wasted in these search methods.

We use a function $r(\ )$, called a ranking function to identify the next answer node speedily. Assigning ranks to live nodes involves some extra "cost" or "computational effort". This extra cost for any node x, denoted by $g(x)$, depends on factors like the number of nodes competing and the number of levels the nearest answer node is in the sub tree of node x.

Suppose an estimate of this extra cost needed to reach an answer node from a node x. This estimate is denoted by $E(g(x))$. Then x is assigned the rank function r such that $E(r(x)) = f(h(x)) + E(g(x))$, where $h(x)$ is the cost of reaching x from the root and f is a non-decreasing function. A search that minimizes $E(r(x))$ is called "Least-Cost" search (LC search). We generate the children of an E-node, and among these, we select the node having least cost function. This method of assigning rank function to competing live nodes makes the searching faster. We now illustrate this method with an example.

Example 8.1

Consider the weighted directed graph of Figure 8.1.

We take initially node 1 as *E*-node and its ranking function is 0 (zero). Generate nodes 2, 3 and 4, the children of node 1. We select node 2 as its ranking function is minimum of $(0 + 2, 0 + 3, 0 + 4) = 2$. Node 2 now becomes *E*-node. Next, generate nodes 5 and 6, children of node 2. We select node 6 as its ranking of function is minimum $(2 + 4, 2 + 1) = 3$. Node 6 now becomes the next *E*-node. Generate nodes 12 and 6, the two children of node 6 (*E*-node). We select node 12 as the *E*-node as its ranking function is minimum of $(3 + 1, 3 + 5) = 4$. The *E*-node 12 has no children. Computation stops. The sequence of the successive *E*-nodes generated is 1, 2, 6 and 12 and this is the LC path.

Consider the control abstraction of the LC-Search method. Let t be a state space tree and $r(\ )$ a cost function for the nodes in t. If x is a node in t, then $r(x)$ is the minimum cost of any answer node in the sub tree with root x. Thus, $r(t)$ is the cost of a minimum-cost answer node in t. As it is very difficult to compute $r(x)$, an heuristic estimate $E(r(x))$ is used. We select this estimate $E(r(x))$ such that if x is either a answer node or a leaf node, then $r(x) = E(r(x))$. In the following LC search algorithm, two functions Least ( ) and Add ( ) are used, to delete and add a live node from or to the list of live nodes, respectively. Least ( ) finds a live node with least $E(r())$. This node is then deleted from the list of live nodes and returned. Add (x) adds the new live node x to the list of live nodes. This list of live nodes will be implemented as a min-heap.

We now give the least cost algorithm.

Algorithm 8.1 LC search

```
Algorithm Clears(rn)
// This algorithm implements the "Least Cost Search".
// If A is an answer node, this algorithm gives the path from
// node A to the root node rn, par(x) gives the parent of node x.
{
   if *rn is an answer node then
   {
      output rn ;
      return ;
   }
   EN := rn ;   // Make rn a E node.
   S: = φ;           //S is the set of all live nodes and is
                        initialized to the empty set;
   repeat
   {
      compute for each child x of EN
      {
         if x is an answer node then
         {
            write (the path from x to rn) ;
            return ;
         }
         Add(x) ;   // x is the required live node. Insert
                        x into the queue.
         par(x) := EN ;
      }
      if all the live nodes are processed then
      {
         write ("No more answer node") ;
         return ;
      }
      EN := Least ( )
   } until (false) ;
}
```

It is to be observed that if the live nodes are represented as a queue and the two functions Least( ) and ADD( ) are used to delete a node from the queue and add a node to the queue, then the LC search becomes a FIFO search. On the other hand, if the live nodes are represented as a stack with the two functions Least( ) and ADD( ) having the above mentioned functions then the LC search becomes a LIFO search.

The way of solving decision problem is through the exhaustive search technique. In this method one generates all candidate solutions and then selects the one(s) with the required property.

Another approach is what is called "backtracking". In this method one solution is constructed at a time, taking a decision at each step. If the partially constructed solution is developed further, as long as it does not violate the problem constraints, then the first legitimate option for the next component is taken. In case, there is no legitimate alternative for the next component or if there are no more remaining options for the next components, then the search stops. To implement this process, we construct a tree of choices called a space state tree.

An improvised technique of the backtracking technique is the branch-and-bound method, and is used for optimization problems. An optimal solution is a feasible solution (a solution that satisfies all the constraints) having the best value of the objective function. Some examples of such problems are the knapsack problem, shortest path problem, the Hamiltonian cycle problem, etc.

In branch-and-bound method, we plan for a method of associating a bound on the best value of the objective function. We associate a lower bound for a minimization problem and an upper bound for a maximization problem.

Before adding a node to the solution that corresponds to the best value of the objective function, which is partially computed up to that stage, we compare the node's bound values with the value of the best solution. If that node is not promising, the branch is terminated.

We terminate a search path at the current node in a state-space tree of a branch-and-bound algorithm for any of the following reasons:

1. The value of the node's bound is not better than the value of the best solution seen so far;
2. The node represents no feasible solution without violating the constraints;
3. When no further choices can be made.

8.3 0/1 KNAPSACK PROBLEM—BRANCH-AND-BOUND

We apply the Least-Cost and Branch-and-Bound to the 0/1 Knapsack problem.

The 0/1 knapsack problem is described earlier. There are n objects and the capacity of the knapsack is m. Profit vector is $(p_1, p_2, \ldots, p_n)$ and the weight vector is $(w_1, w_2, \ldots, w_n)$.

The 0/1 knapsack problem is

$$\text{Maximize} \sum_{i=1}^{n} p_i x_i$$

Subject to $\sum_{i=1}^{n} w_i x_i \le m$, where each $x_i = 0$ or 1 for $1 \le i \le n$.

First construct the state space tree for this problem. We now take the objective function $c(x)$ as $\sum_{i=1}^{n} p_i x_i$, note that this is to be maximized. As our aim is minimization, we take $c(x) = -\sum_{i=1}^{n} p_i x_i$. Those of the leaf nodes, for which $\sum_{i=1}^{n} p_i x_i \le m$ is an answer node, in the sense that it corresponds to a solution $(x_1, x_2, \ldots, x_n)$. We take the cost of the infeasible leaf nodes as $c(x) = \infty$.

For all other internal (non-leaf) nodes, $c(x)$ is defined recursively as min {c(lchild (x)), c(rchild (x))}. We then define two numbers UB(x) and LB(x) (denoting the upper bound and lower bound of the node x), such that $\text{UB}(x) \le c(x) \le \text{LB}(x)$. A cost of $-\sum_{1 \le i < j} p_i x_i$ is made at any node at level j, $1 \le j \le n + 1$.

Therefore, at level j, $c(x) \le q = -\sum_{1 \le i < j} p_i x_i$.

The upper bound number UB(x) is taken as

$$\text{UB}(x) = \text{UBound}\left(q, \sum wx, j-1, m_{ij}\right)$$

where UBound is defined in the following algorithm. LB(x) can be computed using algorithm 7.9 of Chapter 7. It is evident that $c(x) \ge$ Bound $\left(-q, \sum wx, j-1\right)$.

Now we represent an algorithm for computing the UBound function.

Algorithm 8.2 Function u( ) of the Knapsack problem

```
Algorithm UBound (tcp, tcw, rn, m)
//tcp represents the total current profit.
//tcw represents the total current weight.
// The index for the last removed node is rn
//m is the knapsack capacity.
{
    t1 := trp ;
```

```
    t2 := trw ;
    for k = rn + 1 to n do
    {
       if (t2 + w[k] £ m) then
       {
          t2 : = t2 + w[k];
          t1: = t1 - p[k];
       }
    }
    return t1;
}
```

Example 8.2

(LCBB 0/1 knapsack problem): Consider the knapsack problem:

$n = 4$, $(p_1, p_2, p_3, p_4) = (14, 10, 12, 21)$, $(w_1, w_2, w_3, w_4) = (2, 5, 6, 7)$, $m = 15$.

The state space tree of the problem is given in Figure 8.2. The two bounds for nodes 1, 2 and 3 are calculated and are put by the side of these nodes. Computations for other nodes is left as an exercise. The different steps involved in the computation are:

Step 1:

To begin with we put the 1st item in the bag. So, $x_1 = 1$, $w_1 = 2$.

Remaining weight of the bag $= rw = 15 - 2 = 13$.

Weight of the 2nd item $= w_2 = 5$, and can be selected.

Place the 2nd item in the bag. $x_2 = 1$

$$rw = 13 - 5 = 8$$

$$w_3 = 6$$

Place the 3rd in the bag. $x_3 = 1$

$$rw = 8 - 6 = 2.$$

$w_4 = 7$. It cannot be accommodated. $x_4 = 0$.

Solution vector is (1, 1, 1, 0) and the profit $= p_1 + p_2 + p_3 = 14 + 10 + 12 = 36$.

We take UB $= -(p_1 + p_2 + p_3) = -(14 + 10 + 12) = -36$.

Lower bound is calculated by putting fractions of items if necessary.

Lower bound $= -(14 + 10 + 12 + (m - c)/w_4) * p_4)$

$= -(14 + 10 + 12 + (2/7) * 21) = -42$.

We take LB $= -42$.

(In the graph of Figure 8.2, the Upper bound is shown first, followed by the lower bound)

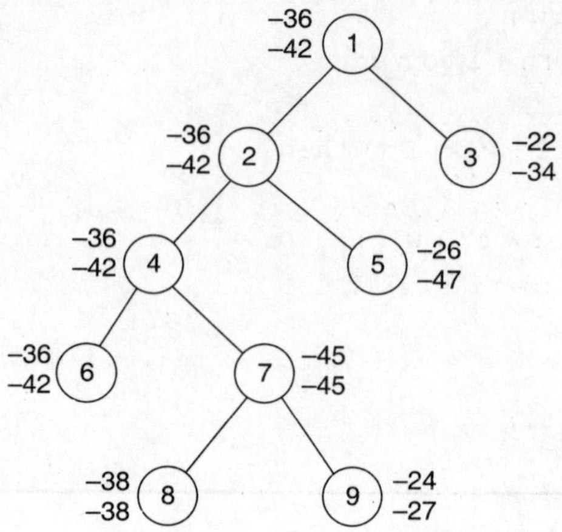

Figure 8.2 LC Branch-and-Bound tree for Example (8.2).

Step II:

Children of node 1 are 2 and 3. Consider these two nodes.

For node 2, $x_1 = 1$

UB = same as above = – 36 and LB = same as above = – 42

The difference UB – LB = – 36 + 42 = 6.

For node 3, $x_1 = 0$

UB = – (0 + 10 + 12) = – 22 and

LB = – (0 + 10 + 12 + ((15 – 11)/7) * 21) = –34.

The difference UB – LB = – 22 + 34 = 12

We select node 2, as it has the difference of bounds which is minimum for it.

Proceeding this way, we calculate all the bounds (shown in Figure 8.2 and the final optimal solution is obtained as (1, 0, 1, 1), i.e. $x_1 = 1$, $x_2 = 0$, $x_3 = 1$, and $x_4 = 1$ and the maximum profit is 47 units.

8.4 TRAVELLING SALESPERSON PROBLEM—LCBB METHOD

We solved the TSP problem using dynamic program approach that has time order complexity of 0 ($n^2\ 2^n$). We now develop a branch-and-bound algorithm for this problem.

$G = (V, E)$ is a weighted directed graph with n vertices that defines an instance of the TSP problem. C_{ij} = cost (or weight) of the edge $(i, j) \in E$. We take $C_{ij} = \infty$ if $(i, j) \notin E$.

Suppose that the tour starts from the node (1) and ends at node (1). No generality is lost in this assumption.

The solution to the state space S is given by $S = \{1, P, 1 \mid P$ is a permutation of $(2, 3, \ldots, n\}$. Note that $|S| = (n - 1)!$

We solve the TSP, using branch-and-bound technique by repeating the following steps:

1. In the computation of cost of the path (i, j), first set all elements of row i and column j to ∞ in the current cost matrix A and also set $A[j, 1]$ to ∞.
2. Apply row, column reduction in A and obtain the total reduction rn.
3. Compute the cumulative cost of the path (i, j) by using the formula:

   Cost (i, j) = Cumulative cost at the previous iteration + $A[i, j] + rn$

Example 8.3

Apply branch-and-bound method to solve the "Travelling Salesperson Problem" whose cost matrix is:

$$\begin{bmatrix} \infty & 18 & 26 & 8 & 11 \\ 14 & \infty & 12 & 7 & 6 \\ 2 & 3 & \infty & 5 & 8 \\ 18 & 4 & 14 & \infty & 7 \\ 15 & 2 & 3 & 19 & \infty \end{bmatrix} \tag{8.1}$$

From each row, the corresponding row minimum is subtracted and for each column the corresponding column minimum is subtracted. Thus, in the above matrix (1), 8, 6, 2, 4 and 2 are subtracted from rows 1, 2, 3, 4 and 5 respectively and then 0, 0, 1, 0 and 0 are subtracted from columns 1, 2, 3, 4 and 5 respectively. This is called row and column reduction. We obtain

$$A = \begin{bmatrix} \infty & 10 & 17 & 0 & 3 \\ 8 & \infty & 5 & 1 & 0 \\ 0 & 1 & \infty & 3 & 6 \\ 14 & 0 & 9 & \infty & 3 \\ 13 & 0 & 0 & 17 & \infty \end{bmatrix} \tag{8.2}$$

Total reduction = 8 + 6 + 2 + 4 + 2 + 1 = 23.

Starting from the node 1, one can visit either of the nodes 2, 3, 4 or 5. We now compute the path $(1, k)$, $2 \le k \le 5$, having a minimum travelling cost.

Path (1, 2):

Set all entries of row 1 and column 2 to ∞ and set $A[2, 1]$ of (2) to ∞ to obtain

$$\begin{bmatrix} \infty & \infty & \infty & \infty & \infty \\ \infty & \infty & 5 & 1 & 0 \\ 0 & \infty & \infty & 3 & 6 \\ 14 & \infty & 9 & \infty & 3 \\ 13 & \infty & 0 & 17 & \infty \end{bmatrix} \tag{8.3}$$

Apply row and column reduction to the above matrix (8.3) and obtain

$$\begin{bmatrix} \infty & \infty & \infty & \infty & \infty \\ \infty & \infty & 5 & 0 & 0 \\ 0 & \infty & \infty & 2 & 6 \\ 11 & \infty & 6 & \infty & 0 \\ 13 & \infty & 0 & 16 & \infty \end{bmatrix} \tag{8.4}$$

Total reduction (tr) = 1 + 3 = 4.
Cost involved in traversing the path (1, 2) is
Cost (1, 2) = 23 + $A[1, 2]$ + tr = 23 + 10 + 4 = 37.

Path (1, 3):

We set all the elements of row 1 and column 3 to ∞ and set A[3, 1] to ∞ in the matrix (8.2) and obtain

$$\begin{bmatrix} \infty & \infty & \infty & \infty & \infty \\ 8 & \infty & \infty & 1 & 0 \\ \infty & 1 & \infty & 3 & 6 \\ 14 & 0 & \infty & \infty & 3 \\ 13 & 0 & \infty & 17 & \infty \end{bmatrix} \tag{8.5}$$

Perform row and column reduction in the above matrix (8.5).

$$\begin{bmatrix} \infty & \infty & \infty & \infty & \infty \\ 0 & \infty & \infty & 1 & 0 \\ \infty & 0 & \infty & 2 & 5 \\ 6 & 0 & \infty & \infty & 3 \\ 5 & 0 & \infty & 17 & \infty \end{bmatrix} \tag{8.6}$$

Total reduction (tr) = 1 + 8 + 1 = 10.
Cost (1, 3) = 23 + $A[1, 3]$ + tr = 23 + 17 + 10 = 50.

Path (1, 4):

Set all the elements of row 1 and column 4 of matrix A of (8.2) and also set $A[4, 1]$ to ∞.

$$\begin{bmatrix} \infty & \infty & \infty & \infty & \infty \\ 8 & \infty & 5 & \infty & 0 \\ 0 & 1 & \infty & \infty & 6 \\ \infty & 0 & 9 & \infty & 3 \\ 13 & 0 & 0 & \infty & \infty \end{bmatrix} \tag{8.7}$$

Perform row and column reduction in the above matrix (8.7). We observe that no reduction need to be done here.

Total reduction = tr = 0.
Cost (1, 4) = 23 + A[1, 4] + tr = 23 + 0 + 0 = 23.

Path (1, 5):

Set all elements of row 1 and column 5 of the matrix A of (8.2) to ∞ and also set A[5, 1] to ∞.

$$\begin{bmatrix} \infty & \infty & \infty & \infty & \infty \\ 8 & \infty & 5 & 1 & \infty \\ 0 & 1 & \infty & 3 & \infty \\ 14 & 0 & 9 & \infty & \infty \\ \infty & 0 & 0 & 17 & \infty \end{bmatrix} \tag{8.8}$$

Perform row and column reduction in the above matrix (8.8).

$$\begin{bmatrix} \infty & \infty & \infty & \infty & \infty \\ 7 & \infty & 4 & 0 & \infty \\ 0 & 1 & \infty & 3 & \infty \\ 14 & 0 & 9 & \infty & \infty \\ \infty & 0 & 0 & 17 & \infty \end{bmatrix} \tag{8.9}$$

Total reduction = tr = 1
Cost (1, 5) = 23 + A[1, 5] + tr = 23 + 3 + 1 = 27.

Thus the cost of paths (1, 2), (1, 3), (1, 4) and (1, 5) are 37, 50, 23, and 27 respectively and the minimum cost corresponds to the path (1, 4), that is the partial optimal path is (1, 2).

$$S = \{1, 4\}$$

One can go from node 4 to any one of the nodes 2, 3 or 5. We now compute the costs of paths (4, 2), (4, 3) and (4, 5). As the minimum cost in the previous iteration corresponds to the path (1, 4), we start the present computations by taking the matrix (8.7).

$$A = \begin{bmatrix} \infty & \infty & \infty & \infty & \infty \\ 8 & \infty & 5 & \infty & 0 \\ 0 & 1 & \infty & \infty & 6 \\ \infty & 0 & 9 & \infty & 3 \\ 13 & 0 & 0 & \infty & \infty \end{bmatrix} \tag{8.10}$$

Path (4, 2):

Set all elements of row 4 and column 2 to ∞ and also set A[2, 4] to ∞, if it is already not ∞.

$$\begin{bmatrix} \infty & \infty & \infty & \infty & \infty \\ 8 & \infty & 5 & \infty & 0 \\ 0 & \infty & \infty & \infty & 6 \\ \infty & \infty & \infty & \infty & \infty \\ 13 & \infty & 0 & \infty & \infty \end{bmatrix} \tag{8.11}$$

We observe that there is no need for any row, column reduction. Therefore,
Total reduction = 0
Cost (4, 2) = 23 + A[4, 2] + tr = 23 + 0 + 0 = 23

Path (4, 3):
Set all the elements of row 4 and column 3 of the matrix (8.10) to ∞ and also set A[3, 1] to ∞.

$$\begin{bmatrix} \infty & \infty & \infty & \infty & \infty \\ 8 & \infty & \infty & \infty & 0 \\ \infty & 1 & \infty & \infty & 6 \\ \infty & \infty & \infty & \infty & \infty \\ 13 & 0 & \infty & \infty & \infty \end{bmatrix} \tag{8.12}$$

Perform row and column reductions in (8.12).

$$\begin{bmatrix} \infty & \infty & \infty & \infty & \infty \\ 0 & \infty & \infty & \infty & 0 \\ \infty & 0 & \infty & \infty & 5 \\ \infty & \infty & \infty & \infty & \infty \\ 5 & 0 & \infty & \infty & \infty \end{bmatrix} \tag{8.13}$$

Total reduction = tr = 1 + 8 = 9.
Cost (4, 3) = 23 + A(4, 3) + tr = 23 + 9 + 9 = 41.

Path (4, 5):
Set all the elements of row 4 and column 5 to ∞ and also set A[5, 1] to ∞.

$$\begin{bmatrix} \infty & \infty & \infty & \infty & \infty \\ 8 & \infty & 5 & \infty & \infty \\ 0 & 1 & \infty & \infty & \infty \\ \infty & \infty & \infty & \infty & \infty \\ \infty & 0 & 0 & \infty & \infty \end{bmatrix} \tag{8.14}$$

Perform row, column reduction in the above matrix (8.14).

$$\begin{bmatrix} \infty & \infty & \infty & \infty & \infty \\ 3 & \infty & 0 & \infty & \infty \\ 0 & 1 & \infty & \infty & \infty \\ \infty & \infty & \infty & \infty & \infty \\ 13 & 0 & 0 & \infty & \infty \end{bmatrix} \tag{8.15}$$

Total reduction = 5
Cost (4, 5) = tr = 23 + A[4, 5] + tr = 23 + 3 + 5 = 31.
We thus see that the costs of the paths (4, 2), (4, 3), and (4, 5) are respectively 23, 41 and 31. Least of these is 23 and corresponds to (4, 2). We select the path from node 4 to 2. The partial optimal tour is 1, 4, 2.

$$S = \{1, 4, 2\}.$$

From node 2 one can go to the remaining nodes 3 or 5. We compute costs of the paths (2, 3) and (2, 5). For these computations, we start with the matrix (8.11) of above as

$$\begin{bmatrix} \infty & \infty & \infty & \infty & \infty \\ 8 & \infty & 5 & \infty & 0 \\ 0 & \infty & \infty & \infty & 6 \\ \infty & \infty & \infty & \infty & \infty \\ 13 & \infty & 0 & \infty & \infty \end{bmatrix} \tag{8.16}$$

Path (2, 3):

We set all the elements of row 2 and column 3 of matrix (8.16) to ∞ and also we set A[3, 1] to ∞.

$$\begin{bmatrix} \infty & \infty & \infty & \infty & \infty \\ \infty & \infty & \infty & \infty & \infty \\ \infty & \infty & \infty & \infty & 6 \\ \infty & \infty & \infty & \infty & \infty \\ 13 & \infty & \infty & \infty & \infty \end{bmatrix} \tag{8.17}$$

We now perform row and column reductions in (8.17).

$$\begin{bmatrix} \infty & \infty & \infty & \infty & \infty \\ \infty & \infty & \infty & \infty & \infty \\ \infty & \infty & \infty & \infty & 0 \\ \infty & \infty & \infty & \infty & \infty \\ 0 & \infty & \infty & \infty & \infty \end{bmatrix} \tag{8.17a}$$

Total reduction = tr = 6 + 13 = 21
Cost (2, 3) = 23 + $A[2, 3]$ + tr = 23 + 5 + 21 = 49.

Path (2, 5):

Set all the elements of row 2 and column 5 of matrix (8.16) to ∞ and also set $A[5, 1]$ to ∞.

$$\begin{bmatrix} \infty & \infty & \infty & \infty & \infty \\ \infty & \infty & \infty & \infty & \infty \\ 0 & \infty & \infty & \infty & \infty \\ \infty & \infty & \infty & \infty & \infty \\ \infty & \infty & 0 & \infty & \infty \end{bmatrix} \tag{8.18}$$

We observe that there is no need to perform any row, column reduction in, matrix (8.18). Therefore, tr = 0.

Cost (2, 5) = 23 + $A[2, 5]$ + tr = 23 + 0 + 0 = 23

Thus the costs of the paths (2, 3) and (2, 5) are 49 and 23 respectively and the minimum cost corresponds to the path (2, 5). Thus the partial optimal path is

$$1 \rightarrow 4 \rightarrow 2 \rightarrow 5$$

$$S = \{1, 4, 2, 5\}$$

The only remaining node left is 3. We consider the path (5, 3). For computation of this, we start with the matrix

$$\begin{bmatrix} \infty & \infty & \infty & \infty & \infty \\ \infty & \infty & \infty & \infty & \infty \\ 0 & \infty & \infty & \infty & \infty \\ \infty & \infty & \infty & \infty & \infty \\ \infty & \infty & 0 & \infty & \infty \end{bmatrix} \quad (8.19)$$

We set all the elements of row 5 and column 3 to ∞ and also set A[3, 1] to ∞.

$$\begin{bmatrix} \infty & \infty & \infty & \infty & \infty \\ \infty & \infty & \infty & \infty & \infty \\ \infty & \infty & \infty & \infty & \infty \\ \infty & \infty & \infty & \infty & \infty \\ \infty & \infty & \infty & \infty & \infty \end{bmatrix} \quad (8.19a)$$

No row and column reduction is possible. Therefore

$$\text{tr} = 0$$

Cost (5, 3) = 23 + A[5, 3] + tr = 23 + 0 + 0 = 23.
The partial optimal path is: $1 \rightarrow 4 \rightarrow 2 \rightarrow 5 \rightarrow 3$

$$S = \{1, 4, 2, 5, 3\}$$

From node 3 one returns to the starting node 1. The final optimal path is:

$$1 \rightarrow 4 \rightarrow 2 \rightarrow 5 \rightarrow 3 \rightarrow 1$$

And the minimum cost is: 8 + 4 + 6 + 3 + 2 = 23.

The basic action in implementing branch and bound method is backtracking. The time complexity T(n) can be reduced if some part of branching can be avoided. We illustrate this point be loss:

Consider four objects with their values and weights as shown below.

Object Number:	1	2	3	4
Value:	15	5	8	4
Weight:	5	3	4	2

Capacity of the knapsack = 10.

Note that object 3 is not selected as inclusion of that violates the capacity of the knapsack (8 + 4 = 12 > 10). With a little thinking, it can be observed that branching to the left from the very start does not improve the solution and as such the entire branching to the left can be done with. This way one reduces the number of branching.

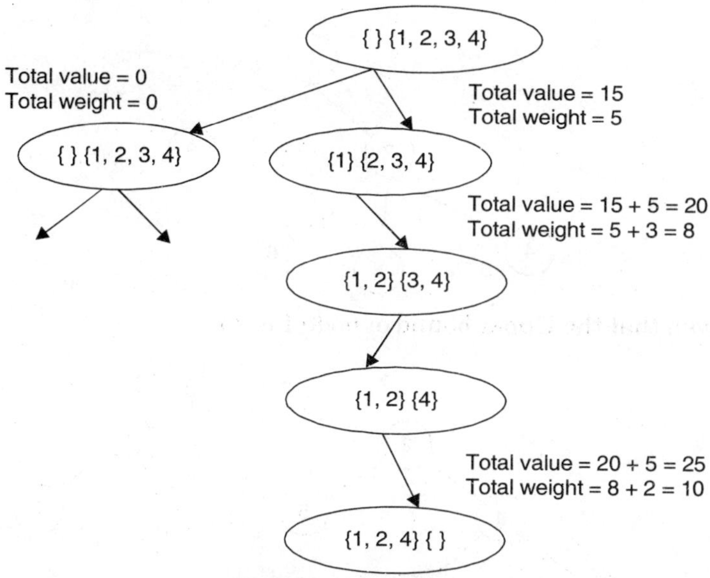

Reduction of branching

QUESTIONS

1. Explain the principle of control abstraction of LC search.
2. Use the LC approach to solve the knapsack problem with $n = 3$, $m = 20$, $(p_1 = 25, p_2 = 24, p_3 = 15)$, and $(w_1, w_2, w_3) = (18, 15, 10)$.
3. What do you mean by "bounding"? Explain how these bounds are useful in branch-and-bound methods.
4. Write an algorithm for the LFOBB for LC answer node.
5. Write an algorithm for the FIFOBB for LC answer node.
6. Explain the LCBB for the 'Travelling salesperson problem'.
7. Draw the portion of the state space tree generated by FIFOBB, LCBB and LIFOBB for the job sequencing with dead lines instance $n = 5$, $(p_1, \ldots, p_5) = (6, 3, 4, 8, 5)$, $(t_1, \ldots, t_5) = (2, 1, 2, 1, 1)$ and $(d_1, \ldots, d_5) = (3, 1, 4, 2, 4)$.
8. Solve the "Travelling Sales Person Problem", using least cost branch-and bound technique, whose cost matrix is:

$$\begin{bmatrix} \infty & 7 & 3 & 12 & 8 \\ 3 & \infty & 6 & 4 & 9 \\ 5 & 8 & \infty & 6 & 18 \\ 9 & 3 & 5 & \infty & 11 \\ 18 & 4 & 9 & 8 & \infty \end{bmatrix}$$

9. Apply FIFOBB to the following problems: given that the Upper Bound UB = 2:

(a)

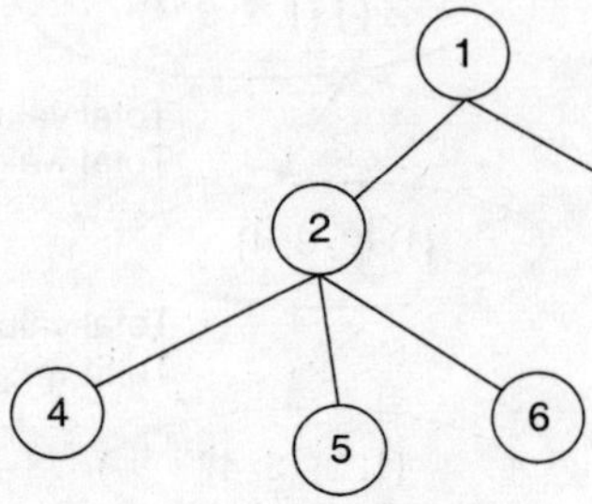

(Given that the Upper bound of node 1 is 2)

(b)

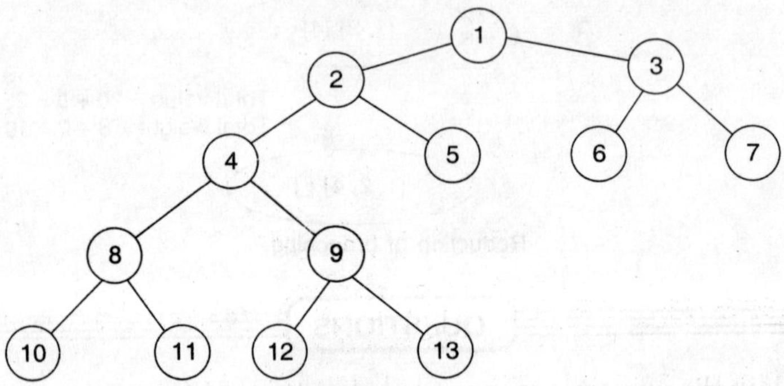

(Given that the upper bound of node 1 is –32)

CHAPTER 9

Algebraic Simplification and Transformations

9.1 INTRODUCTION

Modular arithmetic deals with operations of performing the four basic arithmetic operations modulo p. It is easy and time saving to perform modular arithmetic preferably on integers than the classic arithmetic operations. Necessary algebraic transformations are applied. We begin our study with polynomials. Polynomials play a very important role in algebraic problems. Now consider a way of representing them in the memory of a computer and then consider procedures for performing some operations on them.

We consider a method called "**algebraic transformation**". Suppose we want to compute a function of an input using operators in some non-empty set S_1. Suppose the said computation is very difficult to be done in S_1 from time and efficiency point of view. Then, we transform the input into another form belonging to another set S_2, where in it is easy to compute in S_2, the said function of the input. The input is same in both the sets, but in different forms. Then, we take the "inverse transformation" and come back to the computation of the function in the original set (Think of multiplication of real numbers using logarithms). This process is called "algebraic transformations". A frequently performed algebraic transformation is to convert integers to binary form.

9.2 REPRESENTATION OF POLYNOMIALS

A polynomial, $P(x)$ of degree n (≥ 0) with integer coefficients, is

$$P(x) = a_n x^n + a_{n-1}x^{n-1} + \cdots + a_1 x + a_0$$

Let S_1 be the set of n-degree polynomials. They are represented in set S_2 by the $2n + 1$ pairs $(x_i, P(x_i))$, $1 \le i \le 2n + 1$.

Suppose the function f computes the product of two polynomials $P(x)$ and $Q(x)$, each of degree n.

Computing the product $P(x)\ Q(x)$ directly requires $O(n^2)$ operations. Instead, we consider a transformation of these two polynomials into elements of another set S_2. One way of doing this is by evaluating $P(x)$ and $Q(x)$ at $(2n + 1)$ points. The product is now computed by multiplying the corresponding points. In set S_2, we have the tuples $(x_i, P(x_i)\ Q(x_i))$. Refer to Figure 9.1 for these transformations.

We can determine the product $P(x)\ Q(x)$ in coefficient form by finding the polynomial that interpolates (or satisfies) these $2n + 1$ points.

Example 9.1

Let $n = 3$ and $P(x) = 2x^3 + 4x^2 + 5x + 1$ and $Q(x) = x^3 + x^2 + 2x + 3$. We take $2n + 1 = 7$ points $x_i = 0, 1, 2, 3, 4, 5$, and 6. We compute $P(x_i)$ and $Q(x_i)$ at these seven points.

x_i :	0	1	2	3	4	5	6
$P(x_i)$:	1	12	43	106	213	376	607
$Q(x_i)$:	3	7	19	45	91	163	267
$P(x_i) * Q(x_i)$:	3	84	817	4770	19383	61288	162069

The product of the above two polynomials is represented by the pairs: (0, 3), (1, 84), (2, 817), (3, 4770), (4, 19383), (5, 61288), and (6, 162069).

9.3 POLYNOMIAL EVALUATION

The general form of a **polynomial** of degree n in the variable x is

$$P(x) = a_n\ x^n + a_{n-1}\ x^{n-1} + \cdots + a_1 x + a_0 \tag{9.1}$$

where all a_i are either be integer or floating point constants. a_n is not equal to 0.

A polynomial of type (1) is represented by specifying $(n, a_n, a_{n-1}, \ldots, a_n, a_0)$. This is called the **dense representation**. Some coefficients may be absent. But they are shown as zeros. If more coefficients a_i are absent, the dense representation is waste. Consider the polynomial $x^{100} + 1$. Even though representation requires 102 locations, only three non-zero terms are there. There is a second representation in which only non-zero coefficients and their corresponding exponents are shown. Polynomial of (1), if all $a_i \le 0$ is shown as

$$(n, a_n, n - 1, a_{n-1}, \ldots, n - k, a_{n-k}, \ldots, 1, a_1, 0, a_0) \tag{9.2}$$

If a particular a_k is zero, then k, a_k is not shown in (9.2). This is called the **sparse** representation. The storage saves lot of memory space, particularly when more and more a_i's have zero values.

Suppose we are given the polynomial $P(x) = a_n x^n + \cdots + a_2x^2 + a_1x + a_0$ and we wish to evaluate it at point y. That is why we wish to compute $P(y)$.

The simplest way to evaluate $P(y)$ is to initialize the variable "value" to a_0, then add the term a_ky^k for $1 \le k \le n$ to this value one-term at a time from right to left. The corresponding algorithm is as follows:

Algorithm 9.1 Evaluation of polynomial

```
Algorithm PolyEval (P, n, y)
// This algorithm evaluates a given polynomial P of degree n at a
// specific point y.
{
   mulfac: = 1;
   value: = a0;
   for k: = 1 to n do
   {
      mulfac: = mulfac * y ;
      value : = value + ak * mulfac ;
   }
   return value;
}
```

In the PolyEval algorithm, there are $2n$ multiplications, n additions and $2n + 2$ assignments.

9.3.1 Horner's Rule

The polynomial (1) is rewritten as

$$P(x) = (\cdots ((a_n\, x + a_{n-1})\, x + a_{n-2}\,)\, x + \cdots + a_1)\, x + a_0$$

Following is the corresponding Horner's rule.

Algorithm 9.2 Horner's method

```
Algorithm Horner (P, n, y)
{
      value: = an;
      for k: = n -1 to 0 step -1 do
      {
         value: = value * y + ai ;
      }
      return value;
}
```

Horner's rule requires n multiplication, n additions and $(n + 1)$ assignments. Thus we see that it is an improvement over the straightforward method by a factor of 2.

Suppose the polynomial $P(x)$ has a sparse representation:

$$P(x) = a_m x^{e_m} + \cdots + a_1 x^{e_1}$$

where the $a_i \neq 0$ and $e_m > e_{m-1} > \cdots > e_i \geq 0$. The straightforward algorithm generalizes to the following algorithm to this sparse case.

Algorithm 9.3 Sparse evaluation

```
Algorithm EvalSparse POly (P, m, y )
// Polynomial P is sparse, having m non-zero terms
// with exponents e_i , e_m > e_m-1 > ··· > e_i ≥ 0.
{
    value: = 0.0;
      for k: = 1 to m do
      {
         value : value + a_k * pow (y, e_k);
      }
    return value;
}
```

9.4 INTERPOLATION ALGORITHMS

Given n points (x_i, y_i), we find the unique polynomial $P(x)$ of degree $\leq n - 1$ so that $P(x_i) = y_i$ for $1 \leq i \leq n$. We have two interpolation methods: (1) Lagrange interpolation and (2) Newton's interpolation. In Newton's interpolation scheme, the x values must be equidistant, whereas no such restriction need to be there in Lagrange's method of interpolation.

9.4.1 Lagrange's Interpolation

$$P(x) = \sum_{1 \leq i \leq n} \left[\prod_{i \neq j} \frac{(x - x_i)}{(x_i - x_j)} \right] y_i, \quad 1 \leq j \leq n$$

Degree of $P(x)$ is $\leq n - 1$. It can be verified that $P(x_i) = y_i$, by substituting $x = x_i$ in the above interpolation formula.

Example 9.2

Consider the input (1, 9), (2, 23), and (3, 43). Then the above equation gives

$$P = \frac{(x-2)(x-3)}{(1-2)(1-3)} 9 + \frac{(x-1)(x-3)}{(2-1)(2-3)} 23 + \frac{(x-1)(x-2)}{(3-1)(3-2)} 43$$

$$= 3x^2 + 5x + 1.$$

(Observe that $P(1) = 9$, $P(2) = 23$ and $P(3) = 43$)

Now we represent an algorithm for Lagrange's interpolation scheme.

Algorithm 9.4 Lagrange's interpolation

```
1.  Algorithm Lagrange (x, y, n, LP)
2.  // x and y are one-dimensional arrays each of size n and
3.  // containing the elements (xi) and (yi) respectively, for
    1 ≤ i ≤ n.
4.  // LP is the Lagrange polynomial that interpolates points of
    x, y.
5.  // P is a polynomial.
6.  {
7.       LP: = 0;
8.       for i: = 1 to n do
9.     {
10.        P: = 1;
11.        divfac: = 1;
12.        for j: = 1 to n do
13.        if (i ≠ j ) then
14.        {
15.           P: = P * ( x - x[j] );
16.           divefac: = divfac * ( x[i] - x[j];
17.        }
18.           LP: = LP + ( P * y[i] / divefac);
19.    }
20. }
```

Consider the time complexity of Lagrange's interpolation. Paying attention the multiplication operator and ignoring the subtracting factor in step 16, we note that the time for each execution of the multiplication operator is proportional to the degree of P, which is less than or equal to $j - 1$ at the jth iteration. The total cost of the polynomial multiplication step is

$$\sum_{1\le i\le n} \sum_{1\le j\le n} (j-1) = \sum_{1\le i\le n} \left(\frac{n(n+1)}{2} - n\right)$$
$$= n^2(n+1)/2 - n^2 = 0(n^3)$$

This time complexity is discouraging, because its order is so high.

9.4.2 Newton's Interpolation

We look for another interpolation scheme, called Newton's interpolation method. This method finds the interpolating polynomial by adding one more point (x_{n+1}, y_{n+1}) to the n points (x_i, y_i) points whose interpolating polynomial is known to be $P(x)$.

Suppose

1. $G_{j-1}(x)$ is the interpolating polynomial for $(j-1)$ points (x_k, y_k), $1 \le k < j$, so that $G_{j-1}(x_k) = y_k$ and
2. $D_{j-1}(x) = (x - x_i) \ldots (x - x_{j-1})$.

Then the Newton's interpolation scheme computes $G_j(x)$ by the formula:

$$G_j(x) = [y_j - G_{j-1}(x_j)]\frac{D_{j-1}(x)}{D_{j-1}(x)} + G_{j-1}(x)$$

We observe that

$$G_j(x_k) = [y_j - G_{j-1}(x_j)]\frac{D_{j-1}(x_k)}{D_{j-1}(x_j)} + G_{j-1}(x_k)$$

But $D_{j-1}(x_k) = 0$ for $1 \le k < j$.

So

$$G_j(x_k) = G_{j-1}(x_k) = y_k$$

Then we have

$$\begin{aligned} G_j(x_j) &= [y_j - G_{j-1}(x_j)]\frac{D_{j-1}(x)}{D_{j-1}(x)} + G_{j-1}(x_j) \\ &= y_j - G_{j-1}(x_j) + G_{j-1}(x_j) \\ &= y_j \end{aligned}$$

Example 9.3

Obtain the Lagrange interpolation polynomial for the input data: (0, 6), (2, 12), and (3, 24).

Here $G_1(x) = y_1 = 6$, and $D_1(x) = x - x_1 = x$.

$$\begin{aligned} G_2(x) &= [y_2 - G_1(x_2)]\frac{D_1(x)}{D_1(x_2)} + G_1(x) \\ &= [12 - 6]\, x/2 + 6 \\ &= 3x + 6 \end{aligned}$$

$$\begin{aligned} D_2(x) &= (x - x_1)(x - x_2) \\ &= (x - 0)(x - 2) = x^2 - 2x \end{aligned}$$

Finally,

$$\begin{aligned} G_3(x) &= [y_3 - G_2(x_3)]\frac{D_2(x)}{D_2(x_3)} + G_2(x) \\ &= 3x^2 - 2x + 5 \end{aligned}$$

We now give the necessary algorithm for the above procedure.

Algorithm 9.5 Newton's interpolation

```
Algorithm NewtonInterpolation(X, Y, n, P)
// Assume n ≥ 2. X[1:n] and Y[1:n] are two one-dimensional
// arrays that contain the points: X[i] and Y[i] for 1 ≤ i ≤ n.
// This algorithm returns the interpolating polynomial P
// of degree < n
{
   P: = Y[1];
   Q: = x - X[1];
      for k : = 2 to n do
         {
            dF: = Horner (Q, k-1, X[k]);
            Value of FQL: = Horner (P, k - 2 , X[k]);
            P: = P + (Q*Y[k] - value of Q ) / dF);
            Q: = Q * ( x - X[k] );
         }
}
```

Considering the degrees $(k-1)$ and $(k-2)$ of Q and P at the jth iteration, the total multiplication are

$$\sum_{1 \le j \le n-1} (j-1+j-2) = n(n-1) - 3(n-1) = n^2 - 4n + 3$$

The remaining steps require

$$\sum_{1 \le j \le n-1} (2j) = n(n-1)$$

operations. Thus this interpolating algorithm requires $O(n^2)$ time.

9.5 THE FAST FOURIER TRANSFORM

The fast Fourier transform technique is used to devise efficient algorithms.

The Fourier transform of a continuous function $a(t)$ is given by

$$A(f) = \int_{-\infty}^{\infty} a(t)\, e^{2\pi i f t}\, dt \tag{9.3}$$

The inverse transform of $A(f)$ is

$$a(t) = \frac{1}{2\pi} \int_{-\infty}^{\infty} A(f)\, e^{-2\pi i f t}\, df \tag{9.4}$$

Here, $i = \sqrt{-1}$. The Fourier transform takes a function of time t into a function of frequency f.

As computers deal with discrete systems only, the discrete Fourier transform is defined as follows.

$$A_j = \sum_{0 \le k \le N-1} a_k \, e^{2\pi i \, jk/N}, \quad 0 \le j \le N-1 \tag{9.5}$$

and the inverse is

$$a_k = \frac{1}{N} \sum_{0 \le k \le N-1} A_j \, e^{-2\pi i \, jk/N}, \quad 0 \le j \le N-1 \tag{9.6}$$

In the discrete case, a set $\{a_i\}$ of N sample points is given and a resulting set $\{A_i\}$ of N points is produced for $0 \le i \le N - 1$.

If we substitute $x = w^j$, where $w = e^{2\pi i/N}$ in the following polynomial

$$a(x) = a_{N-1}x^{N-1} + a_{N-2}x^{N-2} + \cdots + a_1x + a_0$$

then we notice that the Fourier element A_j is the value of $a(x)$ at $x = w^j$, for $w = e^{2\pi i/N}$.

Similarly, consider the polynomial with the Fourier coefficients for the inverse Fourier transform.

$$A(x) = A_{N-1}\, x^{N-1} + A_{N-2}\, x^{N-2} + \cdots + A_1x + A_0$$

then each a_k is the value of $A(x)/N$ at $x = (w^{-1})^k$, where $w = e^{2\pi i/N}$.

Thus, we see that discrete Fourier transform is closely connected with polynomial evaluation. The discrete Fourier transform corresponds to the evaluation of a polynomial at N points: $w^0, w^1, \ldots, w^{N-1}$.

We know that we can evaluate a polynomial of nth degree at N points using $0(N^2)$ operations. We apply Horner's rule at each point. The Fast Fourier Transform now turns out to an algorithm for computing these N values using only $0\,(N \log N)$ operations.

9.6 MODULAR ARITHMETIC AND MODULAR TRANSFORMATION

Another algebraic transformation we think of is the modular arithmetic transformation. It basically deals with numbers from a set GF(p) = {0, 1, 2, ..., $p - 1$} for some single precision prime p and as such it is a reformulation of the arithmetic operations.

The mod operator is defined as

$$x \bmod y = x - y\,[x/y] \text{ if } y \ne 0$$

$$x \bmod 0 = x$$

GF(p) forms a field under the following definitions of addition, subtraction, multiplication and division, where $a, b \in$ GF(p):

$$\begin{aligned}(a + b) \bmod p &= a + b && \text{if } a + b < p\\ &= a + b - p && \text{if } a + b \ge p\\ (a - b) \bmod p &= a - b && \text{if } a - b \ge 0\\ &= a - b + p && \text{if } a - b < 0\end{aligned}$$

Note that

$ab \bmod p$ = r such that r is the remainder when the product ab is divided by p;

ab = $qp + r, 0 \le r < \pi$

$a/b \bmod p$ = $ab^{-1} \bmod p = r$, the unique remainder when ab^{-1} is divided by p

ab^{-1} = $qp + r, 0 \le r < p$.

Our aim is to compute the multiplicative inverse b^{-1} of an element $b \in$ GF(p) (Note that b^{-1}) is the multiplicative inverse of b in GF(p) if $bb^{-1} \bmod p = 1$).

For a given $b \in$ GF(p), let $x = b^{-1}$. Then, if there exists an integer y such that $bx = yp + 1$ for some y in GF(p), that is $0 \le k < p$.

For example if $p = 5$

b :	1	2	3	4	5	6	(element)
b^{-1}:	1	4	5	2	3	6	(inverse)
k :	0	1	2	1	2	5	

The multiplicative inverse b^{-1} for an element b of GF(p) – {0} is given by the generalized Euclid's algorithm for the computation of greatest common divisors (gcds). The function gcd (a, b) is defined as

$$\gcd(a,b) = \begin{cases} \gcd(b, a), & \text{if } b > a \\ a, & \text{if } b = 0 \\ \gcd(b, \bmod(a, b)), & \text{if otherwise} \end{cases}$$

where mod (a, b) = a mod b, the remainder on dividing a by b.

Example 9.4

Compute the gcd (20, 8) and (5, 27).

gcd (20, 8) = gcd (8, 4) = gcd (4, 0) = 4.

gcd (5, 27) = gcd (27, 5) = gcd (5, 2) = gcd (2, 1) = gcd (1, 0) = 0

An algorithm for computing gcd (a, b) is given below

Algorithm 9.6 GCD

```
Algorithm gcd(a, b)
// Computes the greatest common divisor of a, b.
{
   gcd(a, 0):= a ;
   return (a);
   if (b > a) then gcd(a, b):= gcd(b, a);
   else
   gcd(a, b): = gcd(b, a mod b);
}
```

The famous Euclid's algorithm establishes the existence of two integers x and y such that $ax + by = \gcd(a, b)$.

An extension of Euclid's algorithm gives the multiplicative inverse b^{-1} of an element $b \in GF(p) - \{0\}$. Let a be a prime p and $b \in GF(p) - \{0\}$. Then $\gcd(p, b) = 1$ and Euclid's algorithm reduces to finding integers x and y such that

$$px + by = 1$$

This implies that y is the multiplicative inverse of b mod p.

Following is an extended Euclidean algorithm.

Algorithm 9.7 Extended algorithm to compute GCD

```
Algorithm ExEuclid (b, p)
//b is in GF(p)-{0}. p is a prime. ExEuclid is a function whose
// result is the integer x such that bx + py = 1.
{
  tp: = p;
  tb: = b;
  x: = 0;
  z = 1;
  while (tb ≠ 1) do
    {
      q: = : ⌊tp /tb⌋;
      t: = tp - tb * q ;
      v: = x - y * q ;
      tp: = tb;
      tb: = t;
      x = y;
      y: = v;
    }
      if y < 0 then y: = y + p;
        return y;
}
```

The computing time of this algorithm is $O(\log_{10} p)$, as the while loop is executed at the most $O(\log_{10} p)$ times.

Modular arithmetic is used as a transformation technique to help us work with integers. Let us represent integers using a set of module, then how we will perform arithmetic on this representation.

Let a and b be integers and suppose that a is represented by the n-tuple $(a_1, a_2, \ldots, a_n)$ where $a_i = a \bmod p_i$ and b is represented by $(b_1, b_2, \ldots, b_n)$ where $b_i = b \bmod p_i$.

The p_i are all single precision primes. This is called a **mixed radix** representation. The rules for addition, subtraction, and multiplication using a mixed radix representation are as follows.

$$\begin{aligned} a + b &= (a_1, a_2, \ldots, a_n) + (b_1, b_2, \ldots, b_n) \\ &= ((a_1 + b_1) \bmod p_1, \ldots, (a_n + b_n) \bmod p_n) \end{aligned}$$

$$a - b = (a_1, a_2, \ldots, a_n) - (b_1, b_2, \ldots, b_n)$$
$$= ((a_1 - b_1) \bmod p_1, \ldots, (a_n - b_n) \bmod p_n)$$
$$ab = (a_1, a_2, \ldots, a_n)(b_1, b_2, \ldots, b_n)$$
$$= ((a_1 b_1) \bmod p_1, \ldots, (a_n b_n) \bmod p_n)$$

Example 9.5

Let $(p_1, p_2, p_3) = (3, 6, 7)$ and let $a = 16$ and $b = 20$.
Then

$$16 = (16 \bmod 3, 16 \bmod 6, 16 \bmod 7) = (1, 4, 2)$$
$$20 = (20 \bmod 3, 20 \bmod 6, 20 \bmod 7) = (2, 2, 6)$$

Therefore

$$16 + 20 = (36 \bmod 3, 36 \bmod 6, 36 \bmod 7) = (0, 0, 1)$$
$$= (1 + 2 \bmod 3, 4 + 2 \bmod 6, 2 + 6 \bmod 7) = (0, 0, 1)$$
$$20 - 16 = (4 \bmod 3, 4 \bmod 6, 4 \bmod 7) = (1, 4, 4)$$
$$= (2 - 1 \bmod 3, 2 - 4 \bmod 6, 6 - 2 \bmod 7) = (1, 4, 4)$$

Also $16 * 20 = (320 \bmod 3, 320 \bmod 6, 320 \bmod 7) = (2, 2, 5)$
$$= ((1 * 2) \bmod 3, (4 * 2) \bmod 6, (2*6) \bmod 7) = (2, 2, 5)$$

Now we need some way of transforming back from modular form with the assurance that the resulting integer is the correct one. Euler proved the following famous theorem in this connection.

Theorem 9.1: (Chinese Remainder Theorem)

Let $n = p_1, p_2, \ldots, p_k$ where n is a positive integer greater than or equal to 2 and $p_1, p_2, \ldots, p_k$ are pairwise relatively prime.Then, there exists a bijection

$$Z_n \rightarrow Zp_1 \times Zp_2 \times \cdots \times Zp_k$$

taken as $a = (a_1, a_2, \ldots, a_k)$, where $a \in Zn_i$, $a_i \in Zp_i$ and $a_i = a \bmod p_i$ for $i = 1, 2, \ldots, k$.

Proof: Define q_i and c_i, for $i = 1, 2, \ldots, k$ as

$$q_i = n/p_i \text{ and}$$
$$c_i = q_i\,(q_i^{-1} \text{ and } p_i),$$

where q_i^{-1} is the multiplicative inverse modulo p_i of q_i.
Compute

$$a \equiv (a_1 c_1 + a_2 c_2 + \cdots + a_k c_k) \bmod n.$$

We claim that

$$a \equiv a_i \bmod p_i$$

Observe that

$$q_j \equiv 0 \bmod p_i \text{ for } j \neq i \text{ and}$$
$$c_i \equiv 1 \bmod p_i.$$

These two facts imply that c is a k-dimensional vector with 1 in its ith position and 0's at all other positions.

$$\begin{aligned} \therefore \quad a &\equiv a_i\, c_i && \bmod p_i \\ &\equiv a_i\, q_i\, (q_i^{-1} \bmod p_i) && \bmod p_i \\ &\equiv a_i && \bmod p_i \end{aligned}$$

This theorem is illustrated by the following example.

Example 9.6

Let $a_1 \equiv 3 \bmod 7$ and $a_2 \equiv 5 \bmod 11$. That is in the notation of the above theorem, $a_1 = 3, a_2 = 5, p_1 = 7, p_2 = 13\ n = 77, q_1 = 13$ and $q_2 = 5$.

$$q_1^{-1} = 11^{-1} \equiv 2 \bmod 7 \text{ and } q_2^{-1} = 7^{-1} \equiv 8 \bmod 11$$

$$\therefore \quad c_1 = 11\,(2 \bmod 7) = 11 * 2 = 22 \text{ and } c_2 = 7\,(8 \bmod 11 = 7 * 8 = 56)$$

$$\begin{aligned} a &= a_1 c_1 + a_2 c_2 \pmod n \\ &= 3 * 22 + 5 * 56 = 346 \pmod{77} \\ &= 16 \pmod{77} \end{aligned}$$

The following algorithm using ExEuclid algorithm and arithmetic modulo p implements the above formula.

Algorithm 9.8 Extended Euclid's algorithm

```
Algorithm CRA (a, p, b, q)
// a and b are elements of in GF(p). gcd (m, n) = 1
// gcd(p, q) = 1.
// This function returns c such that c mod m = a and c mod n = b
{
    r: = a mod q;
    pmq: = p mod qn;
    sol: = ExEuclid (pmq, q);
    v: = ((b - r) * sol) mod q;
    if (v < 0) then v: = v + q;
       r: = v * p + a;
       return r;
}
```

The computing time of the above algorithm is dominated by the call to ExEuclid's that requires $O(\log q)$ operations.

Example 9.7

Compute $5 + 6 * 8$ using modular arithmetic.
We take two primes $p = 7$ and $q = 11$. Then

$$5 = (5 \bmod 7, 5 \bmod 11) = (5, 5)$$
$$6 = (6 \bmod 7, 6 \bmod 11) = (6, 6)$$
$$8 = (8 \bmod 7 , 8 \bmod 11) = (1, 8)$$
$$6 * 8 = (6 * 1 \bmod 7, 6 * 8 \bmod 11) = (6, 4)$$
$$5 + 6 * 8 = ((5 + 6) \bmod 7, (5 + 4) \bmod 11) = (4, 9)$$

Now convert the doublet (4, 9) back to its integer form. We use the CRA algorithm with $a = 4$, $b = 8$, $p = 7$ and $q = 11$.

$$r = a \bmod q = 4 \bmod 11 = 3$$
$$pmq = p \bmod q = 7 \bmod 11 = 7$$
$$\text{sol} = \text{ExEuclid}\ (pmq, q) = \text{ExEulid}\ (7, 11) = 8$$
$$k = 7, p = 11.$$

$(tp, tb, x, y) = (p, k, 0, 1)$ gives

$$tp = 11,\ tb = 7,\ x = 0,\ y = 1$$

Step I:

$$q = \lfloor tp/tb \rfloor = \lfloor 11/7 \rfloor = 1,$$
$$t = tp - tb * q = 11 - 7 * 1 = 4,$$
$$v = x - y * q = 0 - 1 * 1 = -1,$$
$$(tp, tb, x, y) = (tb, t, y, v) \text{ gives}$$
$$tp = 7,\ tb = 4,\ x = 1,\ y = -1$$

This loop is continued till $tb = 1$.

Step II:

From the definitions as given in Step I, we have

$$q = 1,\ t = 3,\ v = 2,\ tp = 4,\ tb = 3,\ x = -1, \text{ and } y = 2$$

As tb $3 \neq 1$, we go to Step III.

Step III:

$$q = 1,\ t = 1,\ v = -3,\ tp = 3,\ tb = 1,\ x = 2,\ y = -3.$$

As $tb = 1$, the iterations stop.

Now, $y \leftarrow y + q = -3 + 11 = 8$ (CRA returns 8).

Is $y < 0$? No. Therefore compute

$$u = (b - r) * \text{sol} \bmod q = (8 - 3) * 8 \bmod 11 = 7.$$
$$r = u * p + a = 7 * 7 + 4 = 53.$$

QUESTIONS

1. Explain Polynomial evaluation and algorithm for the Horner's method.
2. What is interpolation? Write algorithms for (1) Lagrange's interpolation, and (2) Newton's interpolation.
3. Explain "Fast Fourier Transformation".
4. Explain "Modular arithmetic" with the help of some examples.

CHAPTER 10

Lower Bound Theory and NP-Hard and NP-Complete Problems

10.1 INTRODUCTION

Consider a function $g(n)$ that is a lower bound on time complexity of an algorithm ALG1. Suppose that we revise this algorithm and obtain algorithm ALG2. Note that if the computing time of ALG2 is same as $g(n)$, then it is not possible to improve that algorithm.

Recall that if $f(n) = \Omega(g(n))$ then $g(n)$ is a lower bound for $f(n)$. That is, there exist constants c and n_0 such that

$$|f(n)| \geq c\,|g(n)| \quad \text{for all } n > n_0$$

A lower bound related to n (the number of inputs or outputs) exists. For example, for all algorithms that find the maximum of an unordered set of integers, $\Omega(n)$ is a lower bound and for all algorithms for matrix multiplication of two matrices each of order $n \times n$, $\Omega(n^2)$ is a lower bound. For matrix multiplication, the best known algorithm requires $O(n^{2+\epsilon})$ operations for $\epsilon > 0$.

10.2 COMPARISON TREES

Comparison trees are used effectively in deriving lower bounds for certain problems as Searching and Sorting, Merging. In such problems, we make comparisons only and no arithmetic. These are called comparison-based algorithms. Problems such as Search problem which searches a given search element in a given sorted array $A[\ ]$, Sorting problem, Merging problem, are well defined and explained earlier. We now consider them in brief from the comparison-based point of view.

10.2.1 Ordered Searching

This problem is well described by considering a binary tree. A comparison between x and $A[i]$ is associated with each internal node i. Depending on the outcome of this comparison, appropriate decision is taken. If $x = A[i]$, the algorithm terminates successfully. Branching to the left or right of the subtree of the node i is done depending on whether, $x < A[i]$ or $x > A[i]$. If the search reaches a leaf node with $x \neq A[i]$ at any earlier internal node the algorithm terminates unsuccessfully. In this connection, we state the following result.

Result: $A[1 : n]$, $n \geq 1$ is an array of ordered n elements. If FIND(n) is the minimum number of comparison needed, in the worst case, by any comparison based algorithm to recognize whether $x \in A\,[1...n]$, then FIND $(n) \geq \log\,(n + 1)$.

10.2.2 Sorting

Any sorting algorithm can be described again by a comparison tree which is binary tree. Let $A[1\,..\,n]$ be n distinct elements to be sorted. A comparison between $A[i]$ and $A[j]$ is made resulting in the two alternatives: $A[i] < A[j]$ or $A[i] > A[j]$. Thus, each internal node is labeled by the pair (i, j), indicating a comparison between $A[i]$ and $A[j]$. The algorithm terminates on reaching a leaf node. Following figure gives a comparison tree for three nodes.

If $A[i] < A[j]$, the algorithm proceeds down the left branch of the tree otherwise to the right branch. The external nodes represent termination of the algorithm. There is a unique permutation associated with every path from the root to an external node. Following figure gives a comparison tree for three nodes:

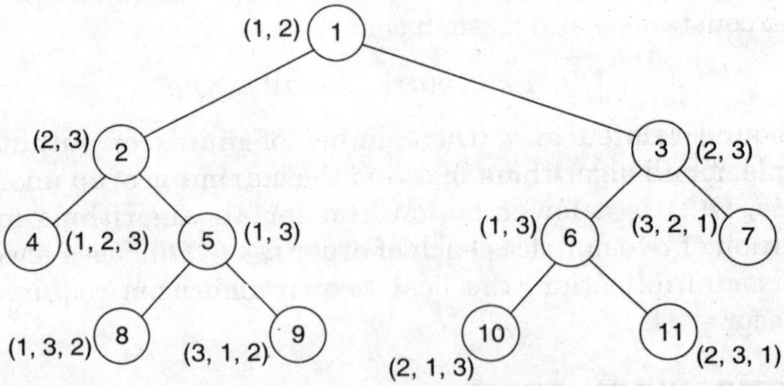

Figure 10.1 A comparison tree for sorting of three elements.

10.3 NP-HARD AND NP-COMPLETE PROBLEMS

All most all algorithms are divided into two groups depending upon their computing times—algorithms having computing times that are bounded by polynomials of small degree and bounded by non-polynomials. Examples of algorithms having computing times of polynomial times are ordered searching ($O(n)$), sorting ($O(n \log n)$), polynomial evaluation ($O(n)$) etc., and travelling sales person ($O(n^2 2^n)$) and knapsack problems ($O(2^{n/2})$) are examples of algorithms having non-polynomial time orders.

Orders of polynomial time algorithms are like $O(\log n)$, $O(n)$, $O(n^2)$, $O(n^3)$, ..., $O(n \log n)$ etc. and the orders of other type of algorithms are: $O(n^2 2^n)$, $O(2^{n/2})$ etc.

We denote the class of decision problems that can be solved by a polynomial time by P.

10.4 NON-DETERMINISTIC ALGORITHMS

One good property of the algorithms we saw is that the output of every operation in them is uniquely defined. Such algorithms are called "**deterministic algorithms**". On the other hand, consider algorithms that are involved in operations whose outcomes are not uniquely defined and that any one of a set of alternatives limited to a termination condition. Such algorithms are called "**non-deterministic algorithms**". Non-deterministic algorithms use the following three functions. S is a set.

1. **Choice (S):** Arbitrarily chooses one of the elements of S.
2. **Failure ( ):** Signals an unsuccessful completion.
3. **Success ( ):** Signals a successful completion.

It is to be noted that the choice(S) function just selects one element from the set S. But, how this choice is made is not specified. If a choice leads to a success, then the algorithm terminates successfully. If the choice leads to a failure, then it selects another element from S and the process is continued. A non-deterministic algorithm terminates unsuccessfully if and only if there exits no set of choices leading to a success. Failure( ) and Success( ) are signals. The computing times of the functions: Choice, Success and Failure are taken each to be $O(1)$.

A machine capable of executing non-deterministic algorithms in this way is called a **non-deterministic machine**. Below some non-deterministic algorithms are given. Observe that the three functions: choice(S), Success( ) and Failure( ) are used to get effective results.

Algorithm 10.1 Non-deterministic search

```
Algorithm NonDetSearch (A, n)
{
k: = choice (1, n);
if A[k] = x then
{
   write (k);
   Success ( );
}
   write (0);
   Failure ( );
}
```

Algorithm 10.2 Non-deterministic sort

```
Algorithm NonDetSort (A, n)
// This is a non-deterministic sorting algorithm.
{
   for j = 1 to n do
      S[j]: = 0;
   for j: = 1 to n do
      {
         k: = choice (1, n);
         if S [k] ≠ 0 then Failure ( );
         else
         S[k] = A [j];
      }
         for j = 1 to n - 1 do
            if S[j] > S[j+1] then Failure ( );
            write (S[1: n]);
            success ( );
}
```

Definition: Any problem for which the answer is either 0 or 1 is called a **decision problem**. The corresponding for a decision problem is called **decision algorithm**. An **optimization problem** is one that identifies an optimal value of a given function (either a maximum value or a minimum value). The algorithm associated with an optimization problem is called an **optimization algorithm**.

Definition: A **maximal complete sub-graph** of a graph $G = (V, E)$ is a clique. The size of the clique is the number of vertices in it. The **max clique problem** is an optimization problem that has to determine the size of the maximum clique in G.

Dclique(G, K) is a deterministic decision algorithm for the clique decision problem.

The size of the max clique in G is known by applying Dclique ( ) repeatedly, for $k = n, n-1, n-2, \ldots,$ until Dclique( ) is 1, n being $|V|$.

Thus, if the time complexity of Dclique is $f(n)$, then the size of a max clique is less than ordered to $nf(n)$. The time for solving a decision problem is same as the time of determining the size of the corresponding max clique. As such the max clique problem can be solved in polynomial time if and only if the clique decision problem can be solved in polynomial time.

The input size of an edge (i, j) is $[\log_2 i] + [\log_2 j] + 2$ in the binary form. The input size of any instance (G, n, k) is

$$n = \sum_{\substack{i<j \\ (i,j)\in G}} \Sigma \lfloor \log_2 i + \log_2 j \rfloor + 2 + [\log_2 k] + 1$$

If G has one connected component then $n \geq |V|$. In view of this, we observe that whether the decision problem can be solved or not with polynomial time complexity $P(|V|)$ depends on whether this problem can be solved or not in a time complexity of $P(n)$.

Example 10.1

Max Clique problem: The following algorithm NonDetClique(G, n, k) is a non-deterministic algorithm for the clique problem. The algorithm begins by forming a set of k distinct vertices. Then the algorithm checks whether these vertices form a complete subgraph. If $G = (V, E)$ is represented by its adjacency matrix and $|V| = n$, the input length m is $n^2 + \lfloor \log_2 k \rfloor + \lfloor \log_2 n \rfloor + 2$. Thus the for loop is implemented in the non-deterministic time $O(n)$. The time for loop of lines 11 and 12 is $O(k^2)$. As such the overall non-deterministic time is $O(n + k^2) = O(n^2) = O(m)$. There is no known polynomial time-deterministic algorithm for this problem.

Algorithm 10.3 Non-deterministic clique pseudocode

```
1.  AlgorithmNonDetClique(G, n, k)
2.  {
3.      Sset: = ϕ;
4.      for i: = 1 to k do
5.          {
6.              ch: = choice (1,n);
7.              if ch ∈ Sset then failure ( );
8.                 Sset: = Sset U {ch}
9.          }
10.             for all pairs ( i, j ) such that i ∈ Sset, j ∈ Sset and i ≠ j do
11.                if ( i, j) is not an edge of G then failure ( );
12.                else
13.                success ( );
14. }
```

Example 10.2

0/1 Knapsack Problem. Consider again the Knapsack problem discussed earlier. Object x_i has a weight w_i and the knapsack has a capacity of m. The knapsack decision problem is to determine whether there is a 0/1 assignment of values to x_i $(1 \leq i \leq n)$ such that $\Sigma\, p_i\, x_i \geq r$ and $\Sigma\, w_i\, x_i \leq m$, where r is a given number and p_i, w_i are non-negative numbers.

If the knapsack decision problem cannot be solved in a deterministic polynomial time, then the optimization problem also cannot be solved in polynomial time.

Assuming p_i, w_i, m and r are all integers, the input size for the knapsack decision problem is

$$\sum_{(1 \le i \le n)} [\log_2 p_i + \log_2 w_i] + 2n + [\log_2 m + \log_2 r] + 2$$

Note that $q > n$. The knapsack decision and the optimization problems can be solved in time $P(s)$ for some polynomial $P(\,)$.

Example 10.3

Knapsack Decision Problem. The following algorithm DKP is a non-deterministic polynomial time algorithm for the knapsack decision problem. The time complexity is $O(n)$. If m is the input length using a binary representation, the time is $O(m)$.

Algorithm 10.4 Non-deterministic 0/1 Knapsack problem

```
Algorithm DKP ( p, w, n, m, r, x )
{
  W: = 0;
  P: = 0;
  for j:= 1 to n do
  {
     x[j]: = choice (0,1);
     W: = w + x[j]*w[j];
  P: = P + x[j] * p[j];
  }
  if ((W > m) or (P < r ) then Failure;
  else
  Success;
}
```

10.5 SATISFIABILITY

The **satisfiability** problem is to determine whether a formula in boolean variables, their compliments, containing the two boolean operators V (**or**) and Λ (**and**) is true for some assignment of truth values of variables. For example the formula $(x_1 \Lambda x_2) V x_3$, is true for the assignment $x_1 = T$, $x_2 = F$ and $x_3 = T$. By CNF-satisfiability, we mean satisfiability the problem when the boolean formula is in conjunctive normal form. If a given propositional formula $E(x_1, x_2, ..., x_n)$ is satisfiable, then a polynomial time non-deterministic algorithm that terminates successfully can be designed and vice versa. Following is a non-deterministic time algorithm for satisfiability.

Algorithm 10.5 A non-deterministic algorithm for satisfiability

```
Algorithm DEval (E,n)
// Checks whether a propositional formula E in the Boolean
// variables x1 , x2 ,..., xn is satisfiable.
```

```
{
    for j = 1 to n do // choose a truth value assignment
       x [k]: = choice (T, F); //T and F denote respectively
 // the boolean constants True, False.
       if E x [1]..., x [n] then Success ( );
       else
       Failure ( );
}
```

10.6 CLASSES OF NP-HARD AND NP-COMPLETE PROBLEMS

An algorithm A is of polynomial computing if there exists a polynomial $p()$ such that the computing time of A is $O(p(n))$ for every input of size n.

Definition: Here P denotes the set of all decision problems solvable by deterministic algorithms in polynomial times NP denotes and the set of all decision problems solvable by non-deterministic algorithms in polynomial times.

Note that $P \subseteq NP$, since deterministic algorithms are just a special case of non-deterministic algorithms.

The famous, unsolved problem in computer science is to determine "**whether $P = NP$ or $P \neq$ NP**".

Now the question concerned with NP is: Are there polynomial time-deterministic algorithms that are not yet considered? A large number of computer science experts are actively engaged to find solution on this problem. As such, there is no proof till to say that $P \neq NP$ or $P = NP$. All these lead to the investigation of whether $NP \subseteq P$.

Prof. S. Cook formulated this question as: "Is there any single problem in NP such that if we showed it to be in P, then that would imply $P = NP$?"

10.7 COOK'S THEOREM

Satisfiability is in P if and only if $NP = P$.

Definition: "If L_1 and L_2 be problems, then we say that problem L_1 reduces to L_2 (also written as $L_1 \alpha L_2$) if and only if there is a way to solve L_1 by a deterministic polynomial time algorithm that solves L_2 in polynomial time".

This definition implies that if we have a polynomial time algorithm for L_2, we can solve L_1 in polynomial time. We observe that α is a transitive. That is if $L_1 \alpha L_2$ and $L_2 \alpha L_3$, then $L_1 \alpha L_3$.

Definition: "A problem L is NP-hard if and only if satisfiability reduces to L (satisfiability α L). A problem L is NP-Complete if and only if L is NP-hard and $L \in NP$" (Figure 10.2).

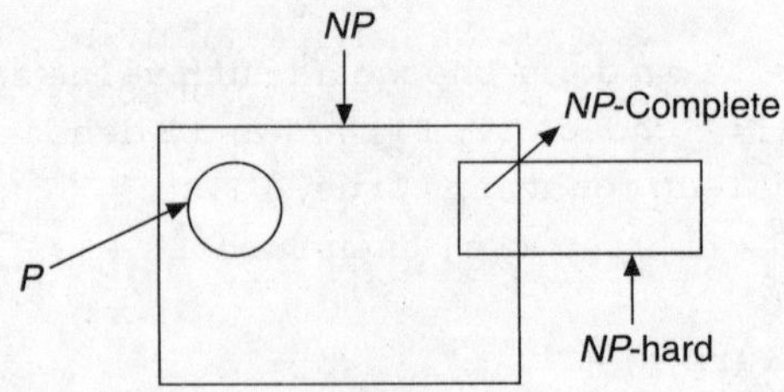

Figure 10.2 A figure of NP-Hard and NP-Complete Problems.

10.8 MORE NP-COMPLETE PROBLEMS

The application of Cook's theorem leads to identifying more problems that are NP-Complete. As a sort of motivation, we consider the following scheduling problem. Let there be n jobs (instructions) $J_1, J_2, \ldots, J_n$ with sizes (execution times) $S_1, S_2, \ldots, S_n$ respectively. That is

Job #:	J_1	J_2	...	J_n
Size:	S_1	S_2	...	S_n

There are only two processors for processing these n jobs. Any job can be processed by any processor (P_1 or P_2). A job can be scheduled to a processor, only if the later is free. Now the problem is to suggest a suitable scheduling scheme so that all the n jobs are processed in the minimum shortest total time possible. There are a number of methods available for this problem. One such scheme (method) is described below.

Take the set S of all the subsets of the set of all jobs. The number of such subsets is 2^n. Take one subset, say A of S. Schedule the jobs in A and obtain $T(A)$, the total time of scheduling the jobs in A. Take the complement A^c of A in S. Schedule all jobs in A^c and obtain $T(A^c)$, the total time of scheduling jobs in A^c. Choose the maximum of $(T(A), T(A^c))$. Repeat this for all other subsets. The optimal scheduling time is then

$$\text{Min}\,\{\text{Max}\,\{T(A), T(A^c)\}\}$$

Suppose you have 1000 jobs. In the brute force method, there are 2^{1000} subsets to be examined. Suppose you can utilize the services of most of the fast computers, connected by a network. Further suppose that all these computers can perform 2^{256} instructions per year. For 1000 jobs, the time you require to carry out all the instructions is

$$2^{1000}/2^{256} = 2^{744} \text{ years}$$

Just imagine how big is this number! Can you solve this problem in your lifetime? If not, the time of how many generations will this problem take to solve?

Such problems are called "**Hard**" problems. **NP-Hard problems are those hard problems that take non-deterministic polynomial times.** In view of this definition, a problem π is said to be **NP-Complete** if

1. $\pi \in$ NP, and
2. π is NP-Hard

Cook's theorem helps in identifying many new problems that are NP-Complete. First, we consider some definitions and list some problems.

10.8.1 Matching in a Graph

A matching in a graph $G = (V, E)$ is a subset M of the edge set E in G such that no two edges in it share the same vertex. In the following graph, the subset $M = \{(a, b), (e, d)\}$ of the edge set is a matching.

10.8.2 Perfect Matching

A perfect matching in a graph $G = (V, E)$ is a matching in G such that each vertex in G is incident on exactly one edge in the matching. Consider the graph in which the subset of edges: $\{(a, b), (c, f), (e, d)\}$ is a perfect matching.

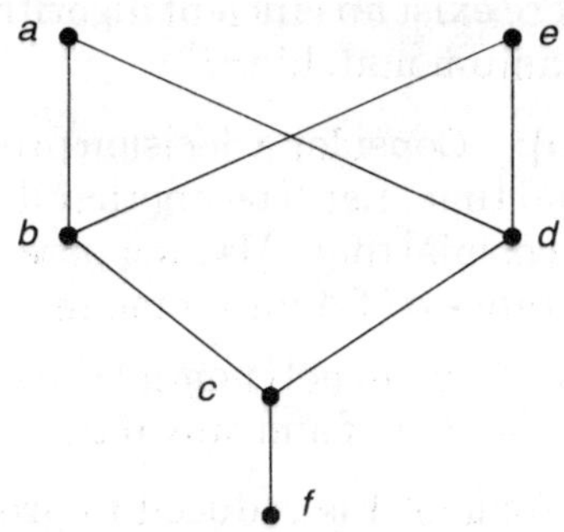

We now take up the following problems and briefly discuss them (without giving the details of their solutions).

Problem 1:

Input: A graph G

Output: Answer to the query "whether G has a perfect matching".

The problems of this type are called decision problems. The output of a decision problem is a single bit (yes or no, otherwise, true or false).

Problem 2:

Input : A graph G.

Output: Obtain a matching of maximum matching.

The problems of this type are called "Search problems". The output of a search problem can have many bits.

Suppose you can solve Problem 2 and obtain a matching of maximum size in a given graph, then the question that we ask is whether "using this solution for Problem 2, can a solution to Problem 1 be obtained".

It can be seen that if the answer is "Yes", then the answer for the reverse problem is also "Yes".

By an efficient algorithm, we mean an algorithm that can be executed successfully on all valid inputs within a reasonable time. Brute force algorithms are not efficient. The following problems are about efficient algorithms.

Problem 3:

Input :	A graph G
Output:	Does there exist an efficient algorithm to check whether G has a perfect matching?

Problem 4:

Input :	A graph G
Output:	Does there exist an algorithm for finding a matching of maximum size?

Problem 5:

Input :	A graph G
Output:	Does there exist an efficient algorithm for finding the size of the maximum matching?

Definition [Reduction]: Consider a decision problem A, which we would like to solve in polynomial time. Let B be another decision problem which is known to be solved in polynomial time. Also, suppose that we have a procedure that transforms any instance α of A into some instance β of B such that

1. Transformation is done in polynomial time, and
2. The answer for α is "Yes" if and only if the answer for β is also "Yes".

Then we say that problem A is reduced to problem B. The method of solving problem B is known. We do not solve problem A directly. Instead, we solve problem B on inputs of problem A and obtain the solution of problem A.

Definition [Hamiltonian paths and cycles]: A Hamiltonian path (HP) in a graph $G = (V, E)$ is a path of length $(n - 1)$. A Hamiltonian cycle (HC) in a graph is a cycle that touches every vertex in G. Consider the following graph:

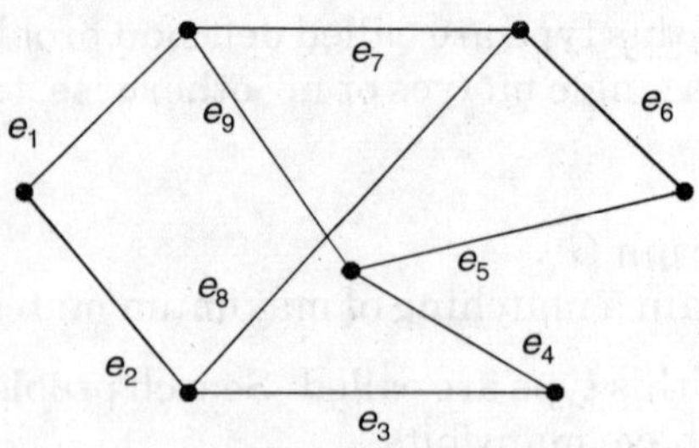

Then, in G, $\{e_1, e_2, e_3, e_4, e_6\}$ is a Hamiltonian path and $\{e_1, e_2, e_3, e_4, e_6, e_7\}$ is a Hamiltonian cycle.

Problem 6

Input: A graph G and an efficient algorithm for HC in G.
Output: Is there an efficient algorithm for a HP in G?

A problem is of **NP class** if:

1. It is a decision problem,
2. It receives "Yes" inputs, and
3. On "Yes" inputs, a proof can be produced for the verification in polynomial time.

A boolean formula (bf) in CNF is an "AND" of clauses connected by " $\wedge$ " only. A clause consists of "OR" of boolean variables or their negations. The value of a boolean formula is true only if every clause is true, otherwise it will be false.

Definition [Satisfiability]: A boolean formula (bf) in CNF is said to be satisfiable if there exists an assignment to the boolean variables involved in the bf such that the bf evaluates to true.

We now look at the following key problem.

Problem 7 (SAT problem)

Input: A boolean formula
Output: Is f satisfiable?

It is proved that the output is "Yes" on "Yes" inputs. This implies that **"The SAT problem belongs to the NP class problems"**

Definition [Clique]: A clique in a graph $G = (V, E)$ is a subset U of the vertex set V such that every pair of vertices in U has an edge in G.

Consider the graph:

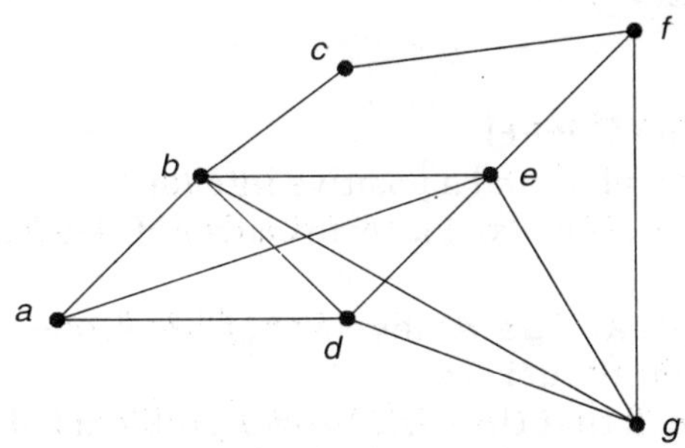

The subset $U = \{a, b, e, d\}$ is a clique in it.

Problem 9 (Clique Search)

Input: A graph G

Output: A clique of the largest size

Problem 10 (Clique Decision)

Input: A graph G and a positive integer k

Output: Does G has a clique of size k?

Problem 11 (Clique-k Decision)

Input: A graph G and a positive integer k.

Output: Does G has a clique of size k?

Definition [Independent Set]: Given a graph $G = (V, E)$, a subset U of V is called an independent set if $\forall\ v_i, v_j$ in V, (v_i, v_j) is not an edge in G (That is, (v_i, v_j) is not in E).

Problem 12 (Independent Set)

Input: A graph G and a positive integer k.

Output: Does G has an independent set of size k?

Definition [Vertex Cover] By a vertex cover of a graph G, we mean a subset U of the vertex set V of G such that every edge in G has at least one end point in U. That is, every vertex in V is incident on at least one edge in G.

In the following graph, $U = \{a, b, c, e, f\}$ is a vertex cover

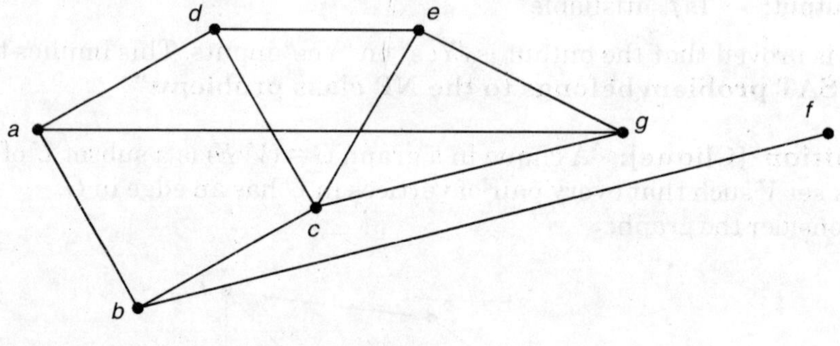

Problem 13 (Vertex Cover)

Input: A graph G and a positive integer k.

Output: Does there exist a vertex cover of size k in G?

Given a graph G having a vertex cover of size k, one can always obtain a vertex cover of the minimum size.

It was established that the vertex cover problem belongs to the class of hard problems. Thus it was established that if there is a polynomial time algorithm for finding a vertex cover of size k, then one can construct a polynomial time algorithm for the independent set and in fact for the clique or for the SAT problems.

Problem 14 (≤ 3 Vertex Cover Problem)

Input: A graph G in which the degree of each vertex is ≤ 3 and a positive integer k.

Output: Does G has a vertex cover of size k?

Problem 15 (Subset Sum Problem)

Input: A set S of n elements with positive sizes $s_1, s_2, \ldots, s_n$ and a positive integer N.

Output: Does there exist a sub set T of S whose size is

$$\text{Size}(T) = \sum_{x \in T} \text{Size}(x) = N$$

The subset sum problem is seen to be NP (Prover/Verifier). One can translate the "XC" problem into a "sum of subsets" problem. Therefore, it is seen that the "sum of subsets" problem is NP-Complete.

Finally, we have the partition problem.

Problem 16 (Partition Problem)

Input: A set T with subsets of T

Output: Is there a partition $\{S, T–S\}$ of T exist such that

$$\text{Size}(S) = \text{Size}(T–S) \qquad (\text{each} = \frac{1}{2}\ \text{Size(T)})$$

It is shown that, if there exists a polynomial time algorithm for the "Partition" problem, then there exists a polynomial time algorithm for the "Sum of subsets" problem. Thus it is seen that the "partition" problem is NP-Complete.

Beginning with Cook's theorem

"SAT problem is NP-Complete"

and using the "reduction" procedure, we see that each of the following results follows:

Cook's theorem asserts that

SAT is NP-Complete

From this, we conclude the following, using the reduction technique:

Clique problem is NP-Complete
Independent set problem is NP-Complete
VC problem is NP-Complete
≤ 3 VC problem is NP-Complete
XC problem is NP-Complete
Subset sum problem is NP-Complete, and finally
Partition problem is NP-Complete.

We feed the inputs into the "Partition" problem and solve the original scheduling problem of n jobs using the two processes.

Example 10.4

(The Halting Problem) The halting problem is to determine whether deterministic algorithm A with an input I ever terminates.

This problem is unsolved.

QUESTIONS

1. Write algorithms for the non-deterministic (1) search, (2) Sort, (3) 0/1 knapsack problems.
2. Differentiate between NP-Hard and NP-Complete problems.
3. Compare and contrast the P and NP problems.
4. State COOK's theorem and explain its importance.
5. What is "halting problem"? Prove that the halting problem is a NP-Hard problem, but not a NP-Complete problem.

References

Aggarwal, Alok and Jeffery Scott Vitter, 'The input/output complexity of sorting and related problems,' *Communication of the ACM,* 31(9): 1116–1127, 1988.

Aho, Alfred V., John E. Hopcroft, and Jeffery D. Ullman, *Data Structures and Algorithms,* Addison-Wesley, 1983.

Ahuja, Ravindra K., Kurt Mehlhorn, James B. Orlin, and Robert E. Tarjan, 'Faster Algorithms for the Shortest Path Problem,' *Journal of the ACM,* 37:213–223, 1990.

Basse, Sara and Alan Van Gelder, *Computer Algorithms: Introduction to Design and Analysis,* 3rd ed., Addison-Wesley, 2000.

Bondey, J.A. and S.R. Murty, *Graph Theory with Applications,* American, Elsevier, 1976.

Brassard, Gilles and Paul Bratley, *Fundamentals of Algorithms,* Prentice Hall, 1988.

Cormen, Thomas H., Charles E. Leiserson, Ronald L. Rivest, and Clifford Stein, *Introduction to Algorithms,* Prentice-Hall of India, 2001.

Horowitz, Ellis, Sartaj Sahni, and Sanguthevar Rajasekaran, *Fundamentals of Computer Algorithms,* Galgotia, 2002.

Knuth, Donald E., *Fundamentals of Algorithms, Vols. 1, 2, 3 of the Art of Computer Programming,* 2nd ed., Addison-Wesley, 1981.

Kozen, Dexter C., *The Design and Analysis of Algorithms,* Springer-Verlag, 1992.

Tremblay, Jean-Paul and Paul G. Sorenson, *An Introduction to Data Structures with Applications,* Tata McGraw-Hill, 1991.

Weis, Mark Allen, *Algorithms, Data Structures and Problem Solving with C++,* Addison-Wesley, Menlo Park, CA, 1996.

Index